# Clostridioides difficile

Henning Sommermeyer • Jacek Piątek

# Clostridioides difficile

## Infections, Risk Factors, Prevention and Treatment

 Springer

Henning Sommermeyer
Department of Health Sciences
Calisia University - Kalisz
Kalisz
Poland

Jacek Piątek
Department of Health Sciences
Calisia University - Kalisz
Kalisz
Poland

ISBN 978-3-030-81099-3        ISBN 978-3-030-81100-6   (eBook)
https://doi.org/10.1007/978-3-030-81100-6



This Springer imprint is published by the registered company Springer Nature Switzerland AG
The registered company address is: Gewerbestrasse 11, 6330 Cham, Switzerland

*This work is dedicated to our wives;
we know that it is not always easy to be
married to a scientist.*

# Foreword

In the last decades, hospital- and community-acquired *Clostridioides difficile* infections (CDI) have become a concern for healthcare worldwide. Broader usage of antibiotics and the introduction of new antibiotics (e.g., certain fluoroquinolones) have resulted in the emergence of hypervirulent and multidrug-resistant strains of *C. difficile*. The global spread of these *C. difficile* strains has resulted in an increased frequency and severity of CDI. These days, *C. difficile* isolated from patients are frequently resistant to some or even several antibiotics. First *C. difficile* strains are found that exhibit a broad range of resistances, including resistance against metronidazole and vancomycin. With the emergence of *C. difficile* strains being resistant against all available antibiotics, CDI is about to become a key challenge for future healthcare.

The book of Sommermeyer and Piątek is providing a comprehensive overview about *C. difficile*, which covers all areas relevant for healthcare professionals, students, and others interested in this subject. Reading the whole book or only individual chapters is providing valuable insights into our current knowledge about the bacterium, its pathophysiology, as well as diagnosis, treatment, and prophylaxis of CDI.

Hanna Pituch
Department of Medical Microbiology
Medical University of Warsaw
Warsaw, Poland

# Preface

Infections caused by *Clostridioides difficile* bacteria are of growing concern for healthcare providers around the world. Triggered by growing usage of old and new antibiotics in healthcare, hypervirulent variants of *C. difficile* strains which are resistant against a broad range of antibiotics have evolved. These hypervirulent strains cause more severe disease and the first strains that are resistant against the remaining effective antibiotics have already been encountered. As *C. difficile* infections are a healthcare-made (antibiotic usage-related) problem, it is imperative for physicians and other healthcare providers to establish a good understanding of the problem and the currently available options to manage it. The present book addresses this need by providing a comprehensive overview of the most recent knowledge in areas of *C. difficile* microbiology, pathophysiology, and epidemiology, as well as risk factors, diagnosis, clinical pictures, and current options of treatment and prophylaxis.

Kalisz, Poland                                                     Henning Sommermeyer

Kalisz, Poland                                                           Jacek Piątek

# Acknowledgments

The authors would like to thank Dr. G. Szymczak from the District Hospital in Jarocin, Poland, for providing the colonoscopy pictures of the chapters "Risk Factors for CDI" and "Clinical Picture of CDI."

Special thanks to Sabine Hanna from Cambridge Assessment English for proofreading, English style editing, and her many useful suggestions.

Thanks also to the colleagues from the Calisia University—Kalisz. Every researcher needs a scientific base—we found ours.

We owe the rector of the Calisia University—Kalisz Prof. Andrzej Wojtyla for his interest, encouragement, and support of our scientific work.

# Contents

# About the Authors

**Henning Sommermeyer** studied biochemistry from 1982 to 1987 at the University of Hanover, Germany. From 1987 to 1990, he specialized in immunology and graduated at the Department of Molecular Pharmacology, at the Medical High School of Hanover. From 1990 to 2010, he worked for different pharmaceutical companies in Germany, Portugal, Poland, and the Czech Republic. In 2010, he founded Pronaos s.r.o., a consultancy company. Since 2019, he is a visiting professor at the Department of Health Sciences of the Calisia University Kalisz, Poland. His scientific work is focused on pro- and synbiotics as prophylactics and therapeutics. He authored and coauthored a number of scientific articles in peer-reviewed scientific journals.

**Jacek Piątek** studied from 1979 to 1985 at the faculty of Medicine at the Medical University of Poznań, Poland. He worked from 1985 until 2018 at the Department of Physiology, University of Medical Sciences in Poznań, Poland. From 2011 to 2013, he was a visiting professor at the Institute of Rural Medicine in Lublin. Since 2018, he is a professor at the Calisia University Kalisz, where he was the Dean of the Faculty of Health Sciences for a year. He is now heading the Internal Disease Department of the university. In recent years, his research interest is focused on the gut microbiota, probiotics, and synbiotics. He is the author of a number of scientific publications on this subject in peer reviewed journals.

# Abbreviations

| | |
|---|---|
| 3D | Three dimensional |
| AB | Antibiotic(s) |
| BC | Before Christ |
| bid | Twice a day |
| bp | Base pairs |
| C | Cytosine |
| CA-CDI | Community-acquired *Clostridioides (Clostridium) difficile* infection |
| Ca-DPA | Calcium dipicolonic acid complex |
| CCCNA | Cell culture cytotoxicity neutralization assay |
| CDI | *Clostridioides (Clostridium) difficile* infection |
| CDT | *C. difficile* dimeric toxin |
| cdtA | Gene for subunit a of *C. difficile* dimeric toxin |
| CDTa | Subunit a of *C. difficile* dimeric toxin |
| cdtB | Gene for subunit b of *C. difficile* dimeric toxin |
| CDTb | Subunit b of *C. difficile* dimeric toxin |
| CLE | Cortex lytic enzyme |
| CROP | Combined repetitive oligopeptide |
| CTn | Conjugative transposon |
| DNA | Desoxyribonucleic acid |
| DPA | Dipicolonic acid |
| EIA | Enzyme immune assay |
| ESCMID | European Society of Clinical Microbiology and Infectious Diseases |
| FDA | U.S. Food and Drug Administration |
| FMT | Fecal microbiota transplantation |
| G | Guanine |
| GDH | Glutamate dehydrogenase |
| H2RA | Histamine-2 receptor antagonists |
| HA-CDI | Hospital (or healthcare)-acquired *Clostridioides (Clostridium) difficile* infection |
| IDSA | Infectious Diseases Society of America |
| IgG | Immunoglobulin G |
| IL | Interleukin |
| kDa | Kilodalton |
| MAL | Muramic-δ-lactam |

| | |
|---|---|
| MLST | Multilocus sequence typing |
| MLVA | Multilocus variable number tandem repeat analysis |
| NAAT | Nucleic acid amplification test |
| NAG | N-acetylglucosamine |
| NAM | N-acetylmuramic acid |
| PCR | Polymerase chain reaction |
| PGFE | Pulsed field gel electrophoresis |
| PMC | Pseudomembranous colitis |
| PPI | Proton pump inhibitor |
| qd | Once daily |
| qid | Four times a day |
| REA | Restriction enzyme analysis |
| SASP | Small acid-soluble proteins |
| SHEA | Society for Healthcare Epidemiology of America |
| TC | Toxigenic culture |
| TCD | Toxigenic *C. difficile* |
| tcdA | Gene of *C. difficile* toxin A |
| TcdA | *C. difficile* toxin A protein |
| TcdB | *C. difficile* toxin B protein |
| tcdB | Gene of *C. difficile* toxin B |
| tid | Three times a day |
| WGS | Whole genome sequencing |

# List of Figures

# List of Tables

# Introduction

**1**

In recent decades, *Clostridioides (C.) difficile* (formerly known as *Clostridium difficile*), a Gram-positive, anaerobic, spore-forming bacteria, has become a major problem for healthcare. First described in the 1930s, *C. difficile* infections (CDI) have seen a strong increase during the last three decades. Interestingly enough, this growth has been fueled mainly by the increased usage of broad-spectrum antibiotics. In asymptomatic *C. difficile* carriers, a normal, diverse, and well-balanced gut microbiota protects against overgrowth of the gut by this pathogen ("colonization resistance"). Disrupting this protective barrier (e.g., by antibiotic therapy) can result in an uncontrolled overgrowth of the gut by *C. difficile*. The resulting disease manifestations range from mild diarrhea to fulminant colitis, which in some patients can result in death. Historically, CDI was considered mainly as a problem patients would acquire in hospitals or healthcare facilities (HA-CDI); however, today a significant portion of CDI is community acquired (CA-CDI). Triggered by the introduction of new antibiotics (e.g., gatifloxacin and moxifloxacin), hypervirulent strains of *C. difficile* (e.g., *C. difficile* ribotype R027) have emerged, which since have spread across the globe.

*C. difficile* and CDI are relevant for all practicing physicians independent of their workplace. While encountering a *C. difficile*-related medical challenge previously may have been most common in a hospital setting, it has now also become increasingly common in community-based healthcare settings. Exposure to healthcare is one of the major risk factors of CDI. Lack of preventative activities (e.g., lack of appropriate hygiene measures fighting *C. difficile* spores) as well as certain actions by healthcare providers (e.g., prescribing antibiotics) can both trigger CDI. Being a healthcare-made problem of the healthcare system makes it mandatory for physicians and other healthcare providers to establish a good understanding of the problem and the currently existing possibilities to manage it.

In the following, the authors have aimed to provide a comprehensive overview of the current knowledge about *C. difficile* and CDI. The following topics are covered by individual chapters: microbiology, pathophysiology, epidemiology, risk factors

© The Author(s), under exclusive license to Springer Nature Switzerland AG 2021
H. Sommermeyer, J. Piątek, *Clostridioides difficile*,
https://doi.org/10.1007/978-3-030-81100-6_1

for CDI, diagnosis, clinical pictures, treatment, and prophylaxis. Reading all chapters will provide the reader with a broad understanding of *C. difficile*, the challenges it may cause, and options to manage the problem. We hope that the way we have structured the text will make it also useful for readers who are only interested in certain specific aspects related to *C. difficile*. The authors admit that sometimes the science described might be a challenge; therefore we have added illustrations, wherever possible and adequate, to facilitate the digestion of the information. Readers interested to go beyond what we have described are recommended to make usage of the extensive reference list. Most of the referenced articles can be downloaded from the internet, free of charge.

# Microbiology **2**

## Contents



## 2.1  Taxonomy

*Clostridium difficile* (Fig. 2.1) was first isolated from the stool of a healthy infant by Hall and O'Toole in 1935 [1].

Because of taxonomic differences between this species and other members of the Clostridium genus, the bacillus was recently renamed *Clostridioides difficile* [2, 3]. The actual scientific classification of *C. difficile* is summarized in Table 2.1.

Vegetative *C. difficile* cells are rod-shaped, pleomorphic, and occur in pairs or short chains. Under the microscope they appear as long, irregular (often drumstick- or spindle-shaped) cells with a bulge at their terminal ends which forms subterminal spores. Under Gram-staining, *C. difficile* cells are Gram-positive and show optimum growth on blood agar at human body temperatures in the absence of oxygen. *C. difficile* is catalase- and superoxide dismutase-negative. The bacteria can produce three types of toxins: enterotoxin A (TcdA), cytotoxin B (TcdB), and *C. difficile* transferase (CDT, or binary toxin). Under stress conditions, *C. difficile* produces spores that are able to tolerate extreme conditions that the vegetative bacteria cells cannot tolerate. *C. difficile* is transmitted by the fecal-oral route and is widely present in the environment. In the last decade, the frequency and severity of *C. difficile* infection has been increasing worldwide to become one of the most common hospital-acquired infections [4].


© The Author(s), under exclusive license to Springer Nature Switzerland AG 2021      3
H. Sommermeyer, J. Piątek, *Clostridioides difficile*,
https://doi.org/10.1007/978-3-030-81100-6_2

**Fig. 2.1** Image of *Clostridioides difficile*

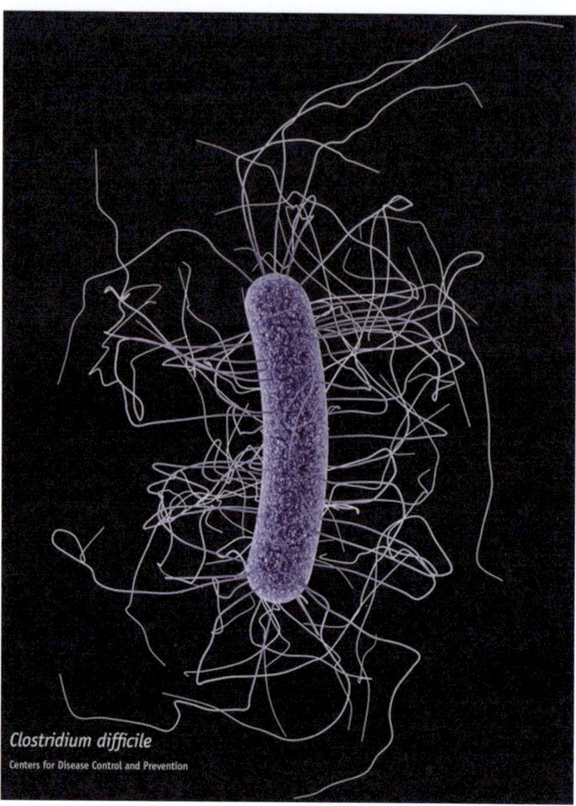


**Table 2.1** Scientific classification of *C. difficile*

| Scientific classification | |
| --- | --- |
| Domain | Bacteria |
| Phylum | Firmicutes |
| Class | Clostridia |
| Order | Clostridiales |
| Family | Peptostreptococcaceae |
| Genus | Clostridioides |
| Species | Clostridioides (C.) difficile |

## 2.2    Genome

In 2006, the first genome sequence of a *C. difficile* strain was published [5]. This multiresistant strain, designated 630, was isolated from a patient with severe pseudomembranous colitis and caused an outbreak of diarrheal disease in a Swiss hospital [6]. The genome consists of a circular chromosome of 4,290,252 base pairs (pb) with a G + C content of 29.06%, and a circular plasmid of 7881 bp with a G + C content of 27.9% [7]. A large proportion (11%) of the genome consists of mobile genetic elements, mainly in the form of conjugative transposons. The

**Table 2.2**  List of completely sequenced *C. difficile* strains

| *C. difficile* strain | Country of origin | Year of isolation | Ribotype |
| --- | --- | --- | --- |
| 196 | France | 1985 | 027 |
| BI1 | USA | 1988 | 027 |
| CF5 | Belgium | 1995 | 017 |
| M68 | Ireland | 2006 | 017 |
| SM (R20291) | UK | 2006 | 027 |
| M120 | UK | 2007 | 078 |
| 855 | USA | 2007 | 027 |

majority of the transposons are conjugative transposons of the Tn916 and Tn1549 families called CTns, which have the ability to excise from their genomic target sites and transpose intra- or intercellular [5, 8]. Exchange of mobile elements occurs frequently and contributes to the plasticity of the genome of *C. difficile* [5, 9, 10]. The mobile elements are putatively responsible for the acquisition by *C. difficile* of an extensive array of genes involved in antimicrobial resistance, virulence, host interaction, and the production of surface structures [11–14]. The metabolic capabilities encoded in the genome show multiple adaptations for survival and growth within the gut environment.

The Sanger Institute has completed sequencing and assembly of several further strains of *C. difficile* (Table 2.2) [7].

## 2.3   Epigenome

An epigenome consists of a record of the chemical changes to the DNA and histone proteins of an organism. These changes can be passed down to an organism's offspring via transgenerational epigenetic inheritance. Changes to the epigenome can result in changes to the structure of chromatin and changes to the function of the genome [15]. The epigenome is involved in regulating gene expression, development, tissue differentiation, and suppression of transposable elements. Unlike the underlying genome, which remains largely static within an individual, the epigenome can be dynamically altered by environmental conditions.

*C. difficile* has a highly diverse epigenome, with 17 high-quality methylation motifs reported so far, the majority pertaining to the 6 mA type. Methylation at one of these motifs, CAAAAA, was shown to impact sporulation, a key step in *C. difficile* disease transmission, as well as cell length, biofilm formation, and host colonization [16].

## 2.4   Life Cycle of C. *difficile* During Infection

### 2.4.1   Spore

*C. difficile* is obligatory anaerobe and survives in the environment in the form of oxygen-resistant spores. Because spores are metabolically dormant, *C. difficile* spores are intrinsically resistant to antibiotics [17], attacks from the host's immune

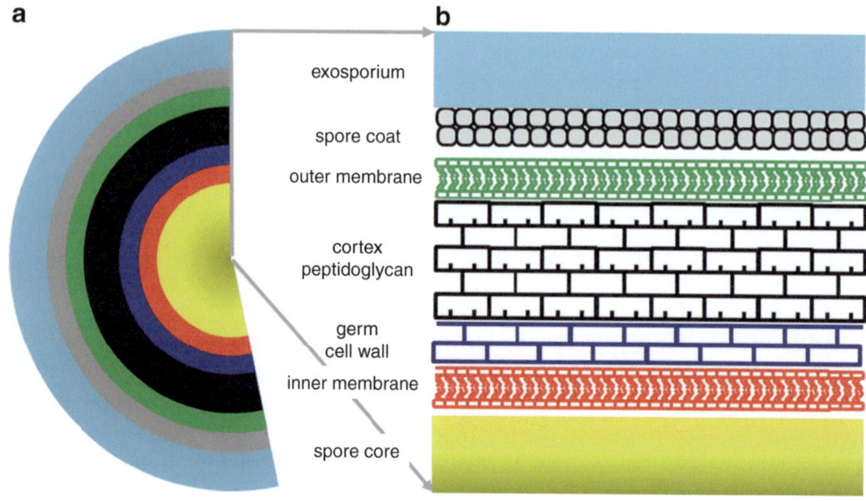



**Fig. 2.2** (**a**) Structural layers of bacterial spores. (**b**) Anatomy of bacterial spore layers without membrane proteins

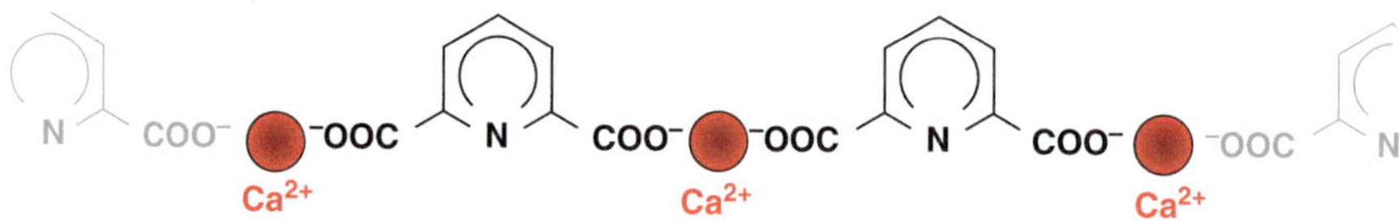

**Fig. 2.3** Calcium-dipicolinate complex (Ca-DPA)

system [18], and, once shed into the environment, they are also resistant to bleach-free disinfectants [19, 20]. The overall structure and morphology of *C. difficile* spores is similar to that of other endospore-forming bacteria [21].

*C. difficile* spores consist of several structural layers (Fig. 2.2a) which contribute to their unique resistance properties. The different layers surrounding the core of the spore have unique properties and compositions (Fig. 2.2b).

### 2.4.1.1 Spore Core

The innermost compartment of the spore is the spore core, which contains the spore DNA, RNA, and most enzymes. The most important factors that contribute to the spore resistance properties are the low water content (25–60% of wet weight), elevated levels of dipicolinic acid (DPA) (25% of the core dry weight) chelated with calcium (Ca-DPA) (Fig. 2.3), and the saturation of DNA with $\alpha/\beta$-type small acid-soluble proteins (SASP) [22].

Dormant spores of *C. difficile* can survive for years, largely because spore DNA is well protected against damage by many different agents. This DNA protection is partly a result of the high level of Ca $2^+$-dipicolinic acid and DNA repair during spore outgrowth, but is primarily caused by the saturation of spore DNA with a group of small acid-soluble spore proteins (SASP) [23]. SASP are synthesized in the developing spore and then degraded after completion of spore germination. The binding of α/β-type SASP to the spore DNA protects against many types of DNA damage, and consequently α/β-type SASP are a major cause of the extreme resistance of spores of *C. difficile* to many different treatments [24, 25].

Of the two SASP types, α/β-type and γ-type, only the α/β-type is expressed in spores of *C. difficile*. SASP are synthesized in the developing spore late in sporulation, and they comprise 8 to 15% of total spore protein and even more of the spore's soluble protein. The α/β-type SASP are products of a multigene family with 4 to 7 genes in *C. difficile*. Although the α/β-type SASP exhibit no sequence similarity to other proteins or protein motifs, these proteins are extremely similar in amino acid sequence, both within and across species. However, even among closely related species there are significant differences among residues in both the C-terminal and N-terminal regions of α/β-type SASP. The likely reason for the high amino acid sequence conservation in α/β-type SASP is that these proteins are nonspecific DNA-binding proteins that saturate the spore DNA. The facts that SASP are such abundant proteins in spores and exhibit sequence differences between closely related species make them good candidates as biomarkers for spore identification [23]. However, effective extraction of these SASP is required.

### 2.4.1.2 Inner Membrane
The spore core is surrounded by a compressed inner membrane [26], which has a similar phospholipid composition as the membrane of the growing bacteria but exhibits very low permeability to small molecules, including water. This feature protects the core from DNA damaging molecules [27, 28].

### 2.4.1.3 Germ Cell Wall
The germ cell wall surrounds the spore inner membrane and becomes the cell wall of the outgrowing bacterium. The germ cell wall consists of peptidoglycan (murein), a polymer consisting of sugars and amino acids (Fig. 2.4) that forms a mesh-like layer outside the plasma membrane of most bacteria.

The sugar component consists of alternating residues of β-[15, 29] linked N-acetylglucosamine (NAG) and N-acetylmuramic acid (NAM). Attached to the N-acetylmuramic acid is a peptide chain of three to five amino acids. The peptide chain can be cross-linked to the peptide chain of another strand, forming the 3D mesh-like layer. Peptidoglycan serves a structural role in the bacterial cell wall, giving structural strength as well as counteracting the osmotic pressure of the cytoplasm.

### 2.4.1.4 Cortex
Surrounding the germ cell wall is the cortex peptidoglycan, whose peptidoglycan expresses three unique modifications: (i) every second muramic acid residue has been converted to muramic-δ-lactam (MAL), a modification not found in the germ cell wall; (ii) about one quarter of cortex N-acetylmuramic acid (NAM) residues are substituted with short peptides providing the cortex with a lower degree of

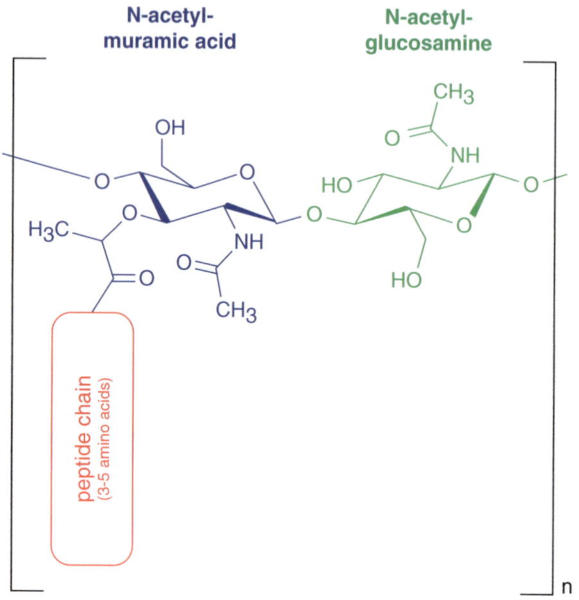

**Fig. 2.4** Chemical structure of peptidoglycan



cross-linking than the germ cell wall; (iii) one fourth of the NAM residues carry a single L-alanine residue, a modification which is absent in the germ cell wall [30, 31]. The MAL residues are specifically recognized by the cortex lytic enzyme (CLE), which hydrolyze the spore cortex, but is not able to degrade the germ cell wall during germination [32].

### 2.4.1.5 Outer Membrane

The cortex peptidoglycan is surrounded by an outer membrane derived from the mother cell during spore formation. This outer membrane confers no resistance properties. The outer membrane contains the enzymes involved in cortex hydrolysis (e.g., the cortex lytic enzyme).

### 2.4.1.6 Coat and Exosporium

A proteinaceous, highly permeable coat surrounds the outer membrane. The spore coat is the outermost layer in some spore-forming bacteria, whereas in others, like in *C. difficile*, an additional layer, the exosporium, is the outermost layer. The exosporium of *C. difficile* of strain 630 has been characterized in detail [33] and a total of 184 proteins were found. Among these were seven spore coat and/or exosporium proteins; six proteins likely to be involved in spore resistance; six proteins possibly involved in pathogenicity; 13 uncharacterized proteins; and 146 cytosolic proteins that might have been encased into the exosporium during assembly. The exosporium is highly permeable and plays a major role in host-pathogen interactions and spore persistence during recurrent infections with *C. difficile*. Table 2.3 provides an overview about the functional roles of the different spore layers.

**Table 2.3** Functional roles of different spore layers

| Spore layer | Physiological role |
|---|---|
| Inner membrane | Permeability barrier protecting against damaging chemicals |
| Germ cell wall | Protection of the inner membrane, will become the cell wall of the vegetative cell |
| Cortex | Degraded by specific cortex lytic enzymes (CLEs) during germination |
| Outer membrane | Contains (CLEs) enzymes involved in cortex hydrolysis |
| Spore coat | Contains (CLEs) enzymes involved in cortex hydrolysis |
| Exosporium | Plays a major role in the host-pathogen interaction |

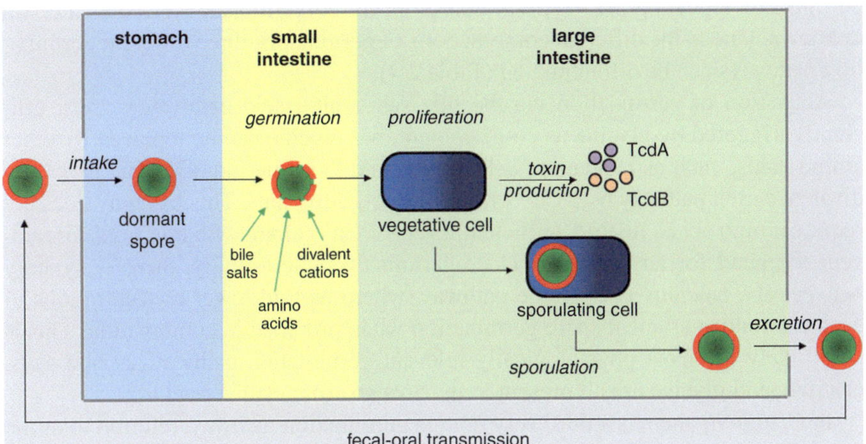


**Fig. 2.5**   Life cycle of *C. difficile* during infection

## 2.4.2   Germination

At the beginning of the life cycle of *C. difficile* [34] during infection (Fig. 2.5), a spore is ingested by the host. Due to its acid-resistant capsule, the spore can pass unharmed through the low pH environment of the stomach.

After arriving in the small intestine, the germination process is triggered normally by nutrients and a variety of non-nutrient agents [35]. In most spore-forming bacteria nutrient germinants are sensed by their cognate germinant receptor (GerA-type germinant receptor), which is localized at the spore's inner membrane [36]. However, *C. difficile* germination is unique, as *C. difficile* does not contain any orthologues of the traditional GerA-type germinant receptor complexes and is the only known spore former to require bile salts in order to germinate [37]. Germination of *C. difficile* is triggered in the gut in response to a combination of bile salts and nutrients present in the gastrointestinal tract, specifically, the bile salt cholate [38]. Most derivatives of cholate can trigger spore germination; however, taurocholate is the most effective germinant in vitro [39–41].

**Table 2.4**  Germination pathways of *C. difficile*

| Pathway | Germinant(s) | Receptors |
| --- | --- | --- |
| Bile salt–amino acid | Taurocholate and glycine, L-histidine, L-serine, or L-alanine | CspC,? |
| Bile salt–divalent cation | Taurocholate and calcium or magnesium | CspC,? |
| Bile salt–alanine racemase-dependent D-amino acid | Taurocholate and D-alanine or D-serine | CspC,? |

Bile salts alone are not sufficient for efficient *C. difficile* germination; a co-germinant, such as amino acids or divalent cations, is essential for the initiation of germination signaling [42–44]. The receptors for co-germinant molecules are still unknown. Due to the different combinations of germinants, three different germination pathways can be differentiated (Table 2.4).

Activation of germination via the bile salt–amino acid pathway is most efficiently triggered by glycine as co-germinant, but glycine can be replaced by other amino acids, such as L-alanine, L-histidine, and L-serine [44–47]. The bile salt–divalent cation pathway uses $Ca^{2+}$ or $Mg^{2+}$ as co-germinants. This pathway does not require amino acids. Instead, sufficient levels of $Ca^{2+}$, along with bile salts, circumvent the need for any amino acid co-germinants. Interestingly, there is synergy between the calcium and glycine pathway, where tenfold-lower concentrations of each can induce efficient spore germination when provided in combination. This is most likely the most physiologically relevant germination pathway, as bile salts, calcium, and glycine are all present in the host gastrointestinal tract [42].

In *C. difficile* there is a third very unique germination pathway, referred to as the alanine racemase-dependent D-amino acid pathway. Alanine racemases typically convert L-alanine to D-alanine (and vice versa). Germination of *C. difficile* spores can be triggered by D-alanine and D-serine in combination with taurocholate, a process for which the alanine racemase Alr2 is required [46].

The sensor for bile salts in the external environment in *C. difficile* is CspC, which is located in the spore coat/outer membrane [48]. Structural CspC resembles a protease; however CspC has no proteolytic activity and has therefore been characterized as a pseudoprotease protein. Stimulation of CspC by taurocholate (or other bile salts) leads to the activation of CspB, which is located in the outer membrane of the *C. difficile* spore. How CspC activates CspB after taurocholate has been detected in the environment remains unknown. CspB is a protease and its activation results in a proteolytic cleavage of pro-SLE, which thereby is transformed into active SleC. For the transformation of pro-SleC into SleC $Ca^{2+}$ is required as a cofactor [42]. SleC is a cortex lytic enzyme (CLE) that will then degrade the spore proteoglycan cortex and stimulate the release of Ca-DPA from the spore core (Fig. 2.6). Absence of CspC, CspB, CspC, or SleC completely blocks germination and outgrowth of *C. difficile* spores [49, 50].

Significant progress has been made during the last years in the understanding of the germination of *C. difficile*. However, there are still some major gaps in our understanding of the germination process details. The role of essential co-germinants

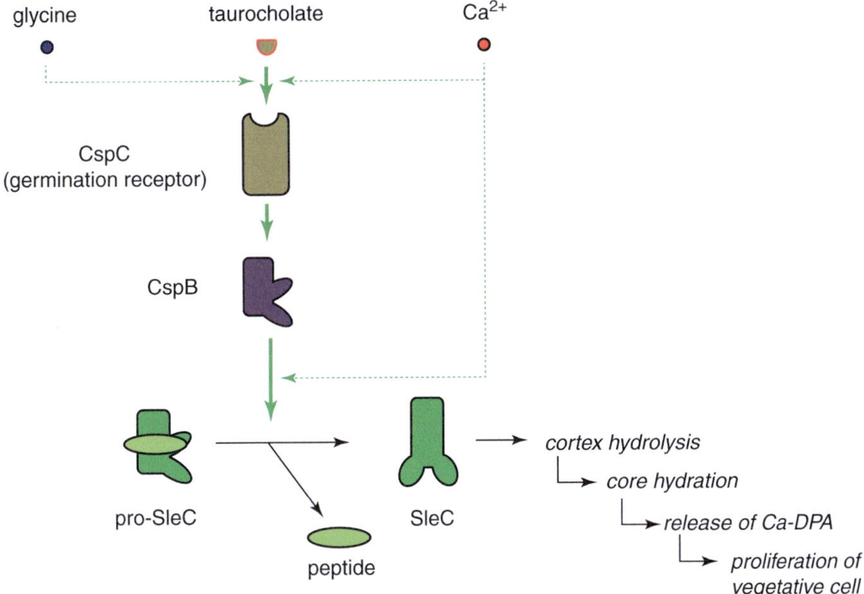


**Fig. 2.6** Overview germination of *C. difficile* spore

and the downstream signaling events that lead to CspB activation are two of these gaps. Currently, two models are discussed for the *C. difficile* germination.

### 2.4.2.1 Germinosome Model

The germinosome model (Fig. 2.7a) assumes that CspC directly activates CspB through some still not characterized protein-protein interaction. CspA is hypothesized to function as a chaperone. Chaperones are proteins which help other proteins to fold correctly. It is assumed that CspA is necessary to allow CspC to incorporate correctly into the mature spore. CspA expression also seems to be controlled, in part, by CspB. Since the activity of each protein encoded by the cspBAC operon appears dependent on others, it is possible that they are forming a "germinosome complex." In this model, after taurocholate binding to CspC and amino acids or/and Ca2+ interacting with the germinosome, CspC activates CspB, which then cleaves the pro-domain from pro-SleC. The cortex lytic enzyme SleC then starts to degrade the cortex.

### 2.4.2.2 Lock-and-Key Model

The lock-and-key model (Fig. 2.7b) hypothesizes that taurocholate binding to CspC is a required event for facilitating the passage of other co-germinants. Thereby taurocholate is a "chemical key" provided by the host, which opens the door for glycine (or other amino acids) and $Ca^{2+}$, allowing them to access deeper layers of the spore. The lock-and-key model addresses two major questions of *C. difficile* germination: (i) how do germinants gain access into the spore and (ii) what role does $Ca^{2+}$

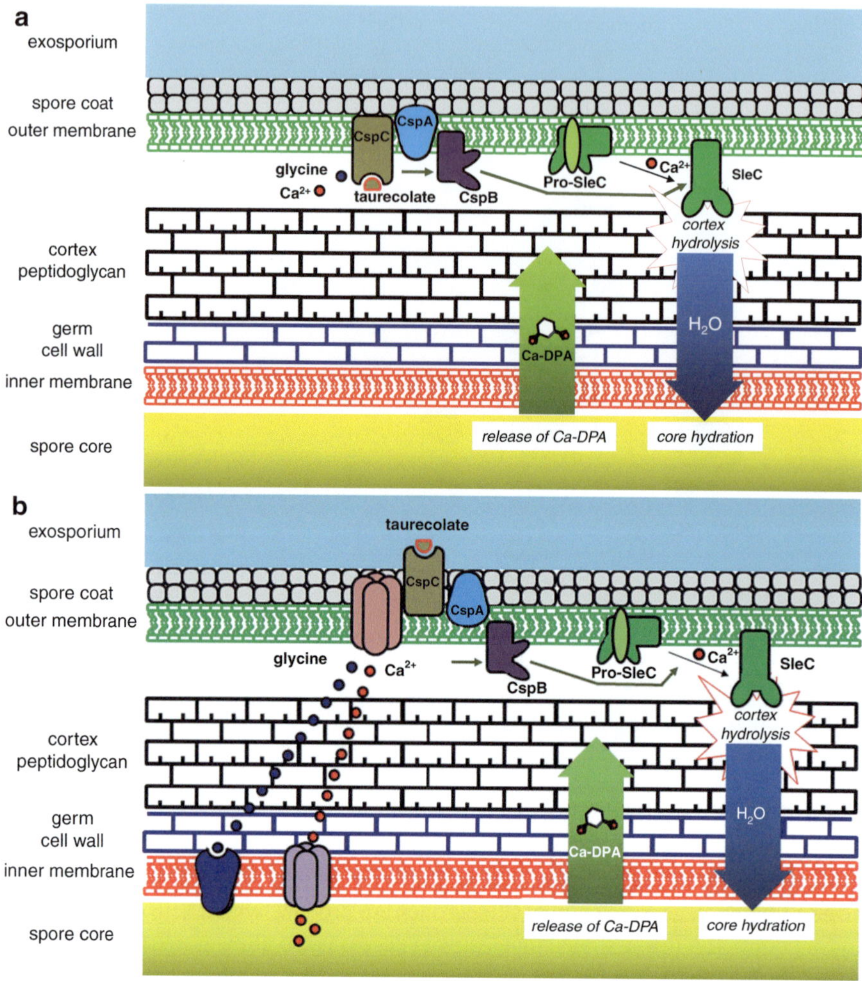



**Fig. 2.7** *C. difficile* germination: (**a**) germinosome model and (**b**) lock-and-key model

play during germination? In this model it is assumed that CspC is outward facing, which enables the protein to detect taurocholate in the environment. The interaction of taurocholate with CspC is assumed to allow co-germinants (via yet unidentified channel proteins?) to enter the inner parts of the spore, where they interact with their putative receptors. These receptors are likely found in the inner membrane, where they facilitate hydration of the core and release of $Ca^{2+}$ and Ca-DPA. $Ca^{2+}$ from the spore core (or the external environment) then binds to and activates CspB which proteolytically cleaves pro-SleC, resulting in the formation of active SleC which then degrades the spore cortex. So far, neither the receptor for amino acids nor that for $Ca^{2+}$ has been identified in the spore of *C. difficile*. The lock-and-key model assumes that $Ca^{2+}$ and amino acids interact with different proteins to facilitate germination; however, it is also possible that $Ca^{2+}$ improves the effectiveness of a single amino acid germinant receptor.

Triggering of the germination is an irreversible process. About 2 h after the germination has been initiated, *C. difficile* will have transformed into a vegetative cell.

### 2.4.3   Sporulation

After the *C. difficile* spore has germinated into a vegetative cell, it will proliferate and outgrow in the gut of the host. Two major events are associated with this phase of the *C. difficile* life cycle: (i) the sporulation (formation of new spores) and (ii) production of toxins, which will be discussed in Chap. 3 of this book.

Dormant spores are produced by vegetative cells of *C. difficile* in the gut of the host. The signals that trigger *C. difficile* sporulation in vivo or in vitro have not yet been identified, but they could be related to environmental stimuli such as nutrient starvation, quorum sensing (the ability to detect and to respond to cell population density by gene regulation), and other unidentified stress factors [51].

In case nutrient limitation is sensed by *C. difficile*, vegetative cells cease to grow and initiate the process of sporulation (Fig. 2.8).

Asymmetric cell division generates a smaller forespore compartment and a larger mother cell [21, 51, 52]. Following completion of DNA segregation, the mother cell proceeds to engulf the forespore. During the initiation of spore metabolic dormancy and compaction of the spore DNA, the mother cell mediates the development of the forespore into the spore through the production of the cortex, outer membrane, coat, and exosporium [52, 53]. Next, the mother cell lyses and the mature spore is released.

In many Bacillus and Clostridium species, the decision to enter sporulation is regulated by several orphan histidine kinases that can phosphorylate the master transcriptional regulator Spo0A [51, 54]. The *C. difficile* strain 630 genome encodes five orphan histidine kinases (CD1352, CD1492, CD1579, CD1949, and CD2492) [55]. Inactivation of D2492 reduces spore formation by 3.5-fold relative to wild type, while a mutation of spo0A completely abolishes spore formation [55]. Although the mechanism by which Spo0A is phosphorylated during the initiation of *C. difficile* sporulation is unclear, it is assumed that the phosphoryl group is transferred directly from the histidine kinases to Spo0A, as has been described for *C. acetobutylicum* [54]. So far, autophosphorylation and transfer of phosphate to Spo0A has only been demonstrated for the *C. difficile* histidine kinase D1579 [55].

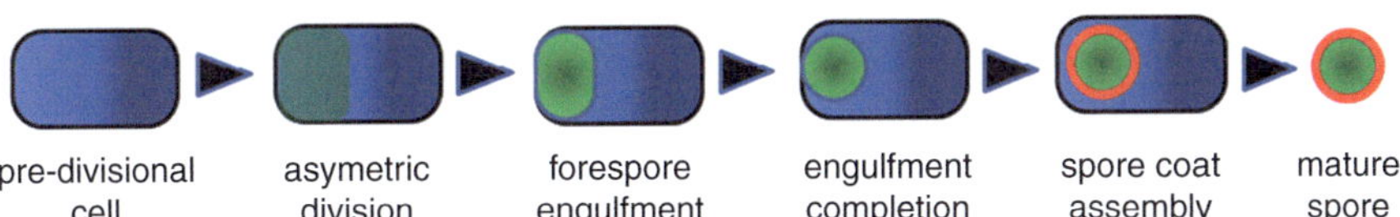


**Fig. 2.8**  Main stages of sporulation

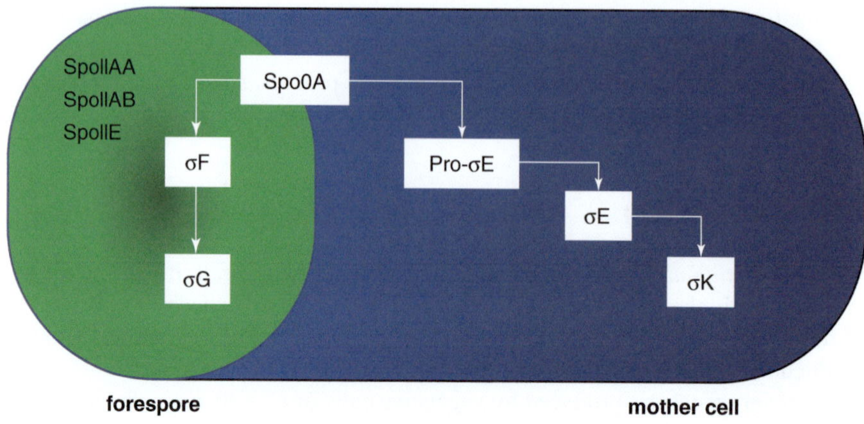



**Fig. 2.9** Model of the global regulatory process of the four main *C. difficile* sporulation-specific sigma factors

Stimulation of the *C. difficile* sporulation pathway results in a cascade of sporulation-specific RNA polymerase sigma factor activations ($\sigma^F$, $\sigma^E$, $\sigma^G$, $\sigma^K$) (Fig. 2.9).

More than 200 genes are regulated by these four sporulation-specific sigma factors. In *C. difficile* $\sigma^F$ regulates 25 genes, $\sigma^E$ 97, $\sigma^G$ 50 and $\sigma^K$ 56 [56, 57]. Spore formation is dependent on the correct orchestration of the process by these RNA polymerase sigma factors. The regulatory pathway controlling sporulation in *C. difficile* is significantly different to those in other Clostridium spp. [29, 56–59].

# References

1. Hall I, O'Toole E. Intestinal flora in newborn infants with a description of a new pathogenic anaerobe, Bacillus difficilis. Am J Dis Child. 1935;49:390.
2. Lawson PA, Citron DM, Tyrrell KL, Finegold SM. Reclassification of Clostridium Difficile as Clostridioides difficile (Hall and O'Toole 1935) Prévot 1938. Anaerobe. 2016;40:95–9. https://doi.org/10.1016/j.anaerobe.2016.06.008.
3. Oren A, Garrity GM. List of new names and new combinations previously effectively, but not validly, published. Int J Syst Evol Microbiol. 2017;67:3140–3. https://doi.org/10.1099/ijsem.0.002278.
4. Czepiel J, Dróżdż M, Pituch H, Kuijper EJ, Perucki W, Mielimonka A, Goldman S, Wultańska D, Garlicki A, Grażyna BG. Clostridium Difficile infection: review. Eur J Clin Microbiol Infect Dis. 2019;38(7):1211–21. https://doi.org/10.1007/s10096-019-03539-6.
5. Sebaihia M, Wren BW, Mullany P, Fairweather NF, Minton N, Stabler R, Thomson NR, Roberts AP, Cerdeño-Tárraga AM, Wang H, Holden MTG, Wright A, Churcher C, Quail MA, Baker S, Nathalie, Brooks K, Chillingworth T, Cronin A, Davis P, Dowd L, Fraser A, Feltwell T, Hance Z, Holroyd S, Jagels K, Moule S, Mungall K, Price C, Rabbinowitsch E, Sharp S, Simmonds M, Stevens K, Unwin L, Whithead S, Dupuy B, Dougan G, Barrell B, Parkhill J. The multidrug-resistant human pathogen Clostridium difficile has a highly mobile, mosaic genome. Nat Genet. 2006;38:779–86. https://doi.org/10.1038/ng1830.
6. Wust J, Sullivan NM, Hardegger U, Wilkins TD. Investigation of an outbreak of antibiotic-associated colitis by various typing methods. J Clin Microbiol. 1982;16:1096–101.

7. Clostridium difficile. 2021. https://www.sanger.ac.uk/resources/downloads/bacteria/clostridium-difficile.html. Accessed 04.06.2021.

8. Brouwer MS, Warburton PJ, Roberts AP, Mullany P, Allan E. Genetic organisation, mobility and predicted functions of genes on integrated, mobile genetic elements in sequenced strains of Clostridium difficile. PLoS One. 2011;6:e23014. https://doi.org/10.1371/journal.pone.0023014.

9. He M, Sebaihia M, Lawley TD, Stabler RA, Dawson LF, Martin MJ, Holt KE, Seth-Smith HMB, Quail MA, Rance R, Brooks K, Churcher C, Harris D, Bentley SD, Burrows C, Clark L, Corton C, Murray V, Rose G, Thurston S, van Tonder A, Walker D, Wren BW, Dougan G, Parkhill J. Evolutionary dynamics of Clostridium difficile over short and long time scales. Proc Natl Acad Sci U S A. 2010;107:7527–32. https://doi.org/10.1073/pnas.0914322107.

10. Stabler RA, Gerding DN, Songer JG, Drudy D, Brazier JS, Trinh HT, et al. Comparative phylogenomics of Clostridium difficile reveals clade specificity and microevolution of hypervirulent strains. J Bacteriol. 2006;188:7297–305. https://doi.org/10.1128/JB.00664-06.

11. Brouwer MS, Roberts AP, Hussain H, Williams RJ, Allan E, Mullany P. Horizontal gene transfer converts non-toxigenic Clostridium difficile strains into toxin producers. Nat Commun. 2013;4:2601. https://doi.org/10.1038/ncomms3601.

12. Jasni AS, Mullany P, Hussain H, Roberts AP. Demonstration of conjugative transposon (Tn5397)-mediated horizontal gene transfer between Clostridium difficile and Enterococcus faecalis. Antimicrob Agents Chemother. 2010;54:4924–6. https://doi.org/10.1128/AAC.00496-10.

13. Mullany P, Wilks M, Lamb I, Clayton C, Wren B, Tabaqchali S. Genetic analysis of a tetracycline resistance element from Clostridium difficile and its conjugal transfer to and from Bacillus subtilis. J Gen Microbiol. 1990;136:1343–9. https://doi.org/10.1099/00221287-136-7-1343.

14. Roberts AP, Mullany P. Tn916-like genetic elements: a diverse group of modular mobile elements conferring antibiotic resistance. FEMS Microbiol Rev. 2011;35:856–71. https://doi.org/10.1111/j.1574-6976.2011.00283.x.

15. Bernstein BE, Meissner A, Lander ES. The mammalian epigenome. Cell. 2007;128(4):669–81. https://doi.org/10.1016/j.cell.2007.01.033.

16. Oliveira PH, Ribis JW, Garrett EM, Trzilova D, Kim A, Sekulovic O, Mead EA, Pak T, Zhu S, Deikus G, Touchon M. Epigenomic characterization of Clostridioides difficile finds a conserved DNA methyltransferase that mediates sporulation and pathogenesis. Nat Microbiol. 2020;5(1):166–80. https://doi.org/10.1038/s41564-019-0613-4.

17. Baines SD, O'Connor R, Saxton K, Freeman J, Wilcox MH. Activity of vancomycin against epidemic Clostridium Difficile strains in a human gut model. J Antimicrob Chemother. 2009;63(3):520–5. https://doi.org/10.1093/jac/dkn502.

18. Paredes-Sabja D, Sarker MR. Interactions between Clostridium perfringens spores and raw 264.7 macrophages. Anaerobe. 2012;18(1):148–56. https://doi.org/10.1016/j.anaerobe.2011.12.019.

19. Ali S, Moore G, Wilson APR. Spread and persistence of Clostridium difficile spores during and after cleaning with sporicidal disinfectants. J Hosp Infect. 2011;79(1):97–8. https://doi.org/10.1016/j.jhin.2011.06.010.

20. Loo VG. Environmental interventions to control Clostridium difficile. Infect Dis Clin N Am. 2015;29(1):83–91. https://doi.org/10.1016/j.idc.2014.11.006. Epub 2015 Jan 5

21. Paredes-Sabja D, Shen A, Sorg JA. Clostridium difficile spore biology: sporulation, germination, and spore structural proteins. Trends Microbiol. 2014;22(7):406–16. https://doi.org/10.1016/j.tim.2014.04.003.

22. Setlow P. I will survive: DNA protection in bacterial spores. Trends Microbiol. 2007;15:172–80. https://doi.org/10.1016/j.tim.2007.02.004.

23. Hathout Y, Setlow B, Cabrera-Martinez RM, Fenselau C, Setlow P. Small, acid-soluble proteins as biomarkers in mass spectrometry analysis of bacillus spores. Appl Environ Microbiol. 2003;69(2):1100–7. https://doi.org/10.1128/AEM.69.2.1100-1107.2003.

24. McDonnell G, Russell AD. Antiseptics and disinfectants: activity, action and resistance. Clin Microbiol Rev. 1999;12:147–79.

25. Nicholson WL, Munakata N, Horneck G, Melosh HJ, Setlow P. Resistance of bacillus endospores to extreme terrestrial and extraterrestrial environments. Microbiol Mol Biol Rev. 2000;64:548–72.

26. Cowan AE, Olivastro EM, Koppel DE, Loshon CA, Setlow B, Setlow P. Lipids in the inner membrane of dormant spores of Bacillus species are largely immobile. PNAS. 2004;101(20):7733–8. https://doi.org/10.1073/pnas.0306859101.

27. Setlow P. Spores of Bacillus Subtilis: their resistance to and killing by radiation, heat and chemicals. J Appl Microbiol. 2006a;101(3):514–25. https://doi.org/10.1111/j.1365-2672.2005 .02736.x.

28. Setlow P. Spores of Bacillus Subtilis: their resistance to and killing by radiation, heat and chemicals. J Appl Microbiol. 2006b;101(3):514–25. https://doi.org/10.1111/j.1365-2672.2005 .02736.x.

29. Al-Hinai MA, Jones SW, Papoutsak ET. σK of Clostridium acetobutylicum is the first known sporulation-specific sigma factor with two developmentally separated roles, one early and one late in sporulation. J Bacteriol. 2014;196(2):287–99. https://doi.org/10.1128/JB.01103-13.

30. Warth AD, Strominger JL. Stucture of the peptidoglycan of bacterial spores: Occurance of the lactam of muramic acid. PNAS. 1969;64(2):528–35. https://doi.org/10.1073/pnas.64.2.528.

31. Warth AD, Strominger JL. Structure of the peptidoglycan from spores of Bacillus Subtilis. Biochemistry. 1972;11(8):1389–96. https://doi.org/10.1021/bi00758a010.

32. Paredes-Sabja D, Sarker MR. Germination response of spores of the pathogenic bacterium clostridium perfringens and Clostridium Difficile to cultured human epithelial cells. Anaerobe. 2011;17(2):78–84. https://doi.org/10.1016/j.anaerobe.2011.02.001.

33. Díaz-González F, Milano M, Olguin-Araneda V, Pizarro-Cerda J, Castro-Córdova P, Tzeng SC, Maier CS, Sarker MR, Paredes-Sabja D. Protein composition of the outermost exosporium-like layer of Clostridium Difficile 630 spores. J Proteome. 2015;123:1–13. https://doi.org/10.1016/j.jprot.2015.03.035.

34. Shen A. A gut odyssey: the impact of the microbiota on Clostridium difficile spore formation and germination. PLoS Pathog. 2015;11(10):e1005157. https://doi.org/10.1371/journal. ppat.1005157.

35. Setlow P. Spore germination. Curr Opin Microbiol. 2003;6(6):550–56. https://doi. org/10.1016/j.mib.2003.10.001.

36. Paredes-Sabja D, Peter Setlow P, Mahfuzur R, Sarker MR. Germination of spores of Bacillales and Clostridiales species: mechanisms and proteins involved. Trends Microbiol. 2011;19(2):85–94. https://doi.org/10.1016/j.tim.2010.10.004.

37. Kochan TJ, Foley MH, Shoshiev MS, Somers MJ, Carlson PE, Hanna PC. Updates to Clostridium difficile spore germination. J Bacteriol. 2018;200(16):e00218–8. https://doi. org/10.1128/JB.00218-18.

38. Sorg JA, Sonenshein AL. Inhibiting the initiation of Clostridium difficile spore germination using analogs of chenodeoxycholic acid, a bile acid. J Bacteriol. 2010;192(19):4983–90. https://doi.org/10.1128/JB.00610-10.

39. Kamiya S, Yamakawa K, Ogura H, Nakamura S. Recovery of spores of Clostridium difficile altered by heat or alkali. J Med Microbiol. 1989;28(3):217–21. https://doi.org/10.109 9/00222615-28-3-217.

40. Weese JS, Staempfli HR, Prescott JF. Isolation of environmental Clostridium Difficile from a veterinary teaching hospital. J Vet Diagn Investig. 2000;12(5):449–52. https://doi. org/10.1177/104063870001200510.

41. Wilson KH, Kennedy MJ, Fekety FR. Use of sodium taurocholate to enhance spore recovery on a medium selective for Clostridium Difficile. J Clin Microbiol. 1982;15(3):443–6.

42. Kochan TJ, Somers MJ, Kaiser AM, Shoshiev MS, Hagan AK, Hastie JL, Giordano NP, Smith AD, Schubert AM, Carlson PA, Hanna PC. Intestinal calcium and bile salts facilitate germination of Clostridium Difficile spores. PLoS Pathog. 2017;13(7):e1006443. https://doi. org/10.1371/journal.ppat.1006443.

43. Shrestha R, Sorg JA. Hierarchical recognition of amino acid co-germinants during Clostridioides difficile spore germination. Anaerobe. 2018;49:41–7. https://doi.org/10.1016/j.anaerobe.2017.12.001.
44. Sorg JA, Sonenshein AL. Bile salts and glycine as cogerminants for Clostridium Difficile spores. J Bacteriol. 2008;190(7):2505–12. https://doi.org/10.1128/JB.01765-07.
45. Ramirez N, Liggins M, Abel-Santos E. Kinetic evidence for the presence of putative germination receptors in Clostridium difficile spores. J Bacteriol. 2010;192(16):4215–22. https://doi.org/10.1128/JB.00488-10.
46. Shrestha R, Lockless SW, Sorg JA. A Clostridium difficile alanine racemase affects spore germination and accommodates serine as a substrate. J Biol Chem. 2017;292(25):10735–42. https://doi.org/10.1074/jbc.M117.791749.
47. Wheeldon LJ, Worthington T, Lambert PA. Histidine acts as a co-germinant with glycine and taurocholate for Clostridium Difficile spores. J Appl Microbiol. 2011;110(4):987–94. https://doi.org/10.1111/j.1365-2672.2011.04953.x.
48. Francis MB, Allen CA, Shrestha R, Sorg JA. Bile acid recognition by the Clostridium difficile germinant receptor, CspC, is important for establishing infection. PLoS Pathog. 2013;9(5):e1003356. https://doi.org/10.1371/journal.ppat.1003356.
49. Burns DA, Heap JT, Minton NP. SleC is essential for germination of Clostridium difficile spores in nutrient-rich medium supplemented with the bile salt taurocholate. J Bacteriol. 2010;192(3):657–64. https://doi.org/10.1128/JB.01209-09.
50. Cartman ST, Minton NP. A mariner-based transposon system for in vivo random mutagenesis of Clostridium difficile. Appl Environ Microbiol. 2010;76(4):1103–9. https://doi.org/10.1128/AEM.02525-09.
51. Higgins D, Dworkin J. Recent progress in Bacillus subtilis sporulation. FEMS Microbiol Rev. 2012;36(1):131–48. https://doi.org/10.1111/j.1574-6976.2011.00310.x.
52. McKenney P, Driks A, Eichenberger P. The Bacillus subtilis endospore: assembly and functions of the multilayered coat. Nat Rev Microbiol. 2013;11:33–44. https://doi.org/10.1038/nrmicro2921.
53. Henriques AO, Moran CP. Structure, assembly, and function of the spore surface layers. Annu Rev Microbiol. 2007;61(1):555–88. https://doi.org/10.1146/annurev.micro.61.080706.093224.
54. Steiner E, Dago AE, Young DI, Heap JT, Minton NP, Hoch JA, Youn M. Multiple orphan histidine kinases interact directly with Spo0A to control the initiation of endospore formation in Clostridium acetobutylicum. Mol Microbiol. 2011;80(3):641–54. https://doi.org/10.1111/j.1365-2958.2011.07608.x.
55. Underwood S, Guan S, Vijayasubhash V, Baines SD, Graham L, Lewis RJ, Wilcox MH, Stephenson K. Characterization of the sporulation initiation pathway of Clostridium difficile and its role in toxin production. J Bacteriol. 2009;191(23):7296–305. https://doi.org/10.1128/JB.00882-09.
56. Fimlaid KA, Bond JP, Schutz KC, Putnam EE, Leung JM, Lawley TD, Shen A. Global analysis of the sporulation pathway of Clostridium difficile. PLoS Genet. 2013;9(8):e1003660. https://doi.org/10.1371/journal.pgen.1003660.
57. Saujet L, Pereira FC, Serrano M, Soutourina O, Monot M, Shelyakin PV, Gelfand MS, Dupuy B, Henriques AO, Martin-Verstraete I. Genome-wide analysis of cell type-specific gene transcription during spore formation in Clostridium difficile. PLoS Genet. 2013;9(10):e1003756. https://doi.org/10.1371/journal.pgen.1003756.
58. De Hoon MJL, Eichenberger P, Vitkup D. Hierarchical evolution of the bacterial sporulation network. Curr Biol. 2010;20(17):R735–45. https://doi.org/10.1016/j.cub.2010.06.031.
59. Harry KH, Zhou R, Kroos L, Melville SB. Sporulation and enterotoxin (CPE) synthesis are controlled by the sporulation-specific sigma factors SigE and SigK in Clostridium perfringens. J Bacteriol. 2009;191(8):2728–42. https://doi.org/10.1128/JB.01839-08.

## Contents



## 3.1    Toxins TcdA and TcdB

Virulence is a pathogen's or microbe's ability to infect or damage a host. In the case of *C. difficile* the production of toxins and their effects on the host's gut epithelium is a major virulence factor [1]. Two major toxins, TcdA and TcdB, are produced by *C. difficile*. Understanding the activities of these toxins provides a good insight into the *C. difficile*-related disease. In addition to their contribution to disease, TcdA and TcdB are the primary markers for diagnosis of *C. difficile* disease and are detected in the stools of patients by antibody-based and cytotoxicity assays.

TcdA (308 kDa) and TcdB (270 kDa) are glycosyltransferases, which inactivate Rho, Rac, and Cdc42 within target cells. TcdA and TcdB are among the largest bacterial toxins reported and belong to the group of large clostridial toxins. TcdA and TcdB are encoded on a pathogenicity region within the chromosome of *C. difficile* and are expressed efficiently during the late log and stationary phases of growth in response to a variety of environmental stimuli. The glycosylating activities of the toxins TcdA and TcdB modulate numerous physiological events in the cell and contribute directly to disease.

Genetic manipulation of *C. difficile* aiming for the generation of isogenic strains deficient in toxin production is difficult. Nevertheless a number of observations support the assumption that the toxins are directly involved in disease manifestation: (i) there is a direct relationship between toxin levels and the development of pseudomembranous colitis and duration of diarrhea [2], (ii) levels of immunoglobulin G against TcdA correlate directly with protection from disease following colonization [3], and (iii) TcdA administration to animals results in the establishing of the hallmarks of pseudomembranous colitis, including fluid accumulation, inflammation, and cell damage [4, 5].

The role of TcdB in disease is not as well understood as the contributions of TcdA. In an early study it was found that TcdB was unable to initiate disease unless TcdA was present [6]. However, TcdA⁻TcdB⁺ strains of *C. difficile* are occasionally identified from clinical isolates and were found to cause disease, in some cases extensive pseudomembranous colitis [7].

## 3.2    Genetics of TcdA and TcdB

The genes encoding TcdA and TcdB, tcdA and tcdB, have been sequenced and are found in a single open reading frame located within a 19.6 kb pathogenicity locus [8, 9]. Both open reading frames are large, with tcdA found within an 8133 nucleotide region and tcdB is 7098 nucleotides in length.

Both tcdA and tcdB are low G + C (<28%) genes, which is comparable to the G + C content (≈29%) of the *C. difficile* genome, and the toxins exhibit a high degree of overall similarity (66%). Given the proximal location of tcdA and tcdB and the high sequence and functional homology between the two proteins, it has been proposed that the two genes may have arisen as the result of a gene duplication event [10]. Furthermore, similarity in the biochemical activity of TcdA and TcdB, wherein both toxins use a highly conserved N-terminal domain to modify identical substrates, support the notion of gene duplication. The major regions of homology between TcdA and TcdB fall within the enzymatic and receptor-binding domains of the two toxins [11]. The N-terminal domains of TcdA and TcdB show 74% homology, and this homology provides a basis for the similar substrate specificity of these two toxins.

The C-termini of TcdA and TcdB show a number of short, homologous regions termed combined repetitive oligopeptides (CROPs) [10]. TcdA encodes five groups of CROPs, which range in size from 21 to 50 residues and can be repeated throughout the C-terminus of the protein. TcdB also encodes five groups of CROPs, four of which show homology to the CROPs of TcdA. Yet the CROPs found in TcdB are more divergent and less frequent than those found in TcdA. CROPs appear to play a putative role in initial target cell interaction and receptor binding, but the mechanism explaining the necessity for these repeats in cell binding remain unclear.

## 3.3   Pathogenicity Locus

TcdA and TcdB are both encoded on the same ≈19.6 kb pathogenicity locus (Fig. 3.1) in *C. difficile* [12].

The two toxin genes are closely situated, with a 1350-nucleotide intervening sequence on this locus, and are transcribed in the same direction. In addition to tcdA and tcdB, three other open reading frames are located on this pathogenicity locus and are thought to be involved in regulation of toxin production or release of the toxin from the cell [12]. The gene tcdC lies downstream of tcdA and is transcribed in the opposite direction from the two toxin genes. Expression of tcdC is high in early exponential phase but declines as growth moves into the stationary phase [13]. The decline in TcdC expression corresponds to increases in TcdA and TcdB, suggesting that TcdC may function as a negative regulator of toxin production.

The gene tcdD is found upstream of tcdB and is coordinately expressed with both of the toxin genes. TcdD is similar to DNA-binding proteins and has been shown experimentally to enhance expression of promotor reporter fusions containing the promoter-binding regions of tcdA and tcdB [14]. Expression of TcdD is responsive to environmental conditions and is repressed by glucose. TcdD expression is significantly increased in the stationary phase [13]. Thus, TcdD appears to be a major positive regulator of tcdA and tcdB expression.

The gene encoding TcdE is positioned between tcdB and tcdA and shows homology with holing proteins and thus has been speculated to facilitate the release of TcdA and TcdB through permeabilization of the *C. difficile* cell wall [15].

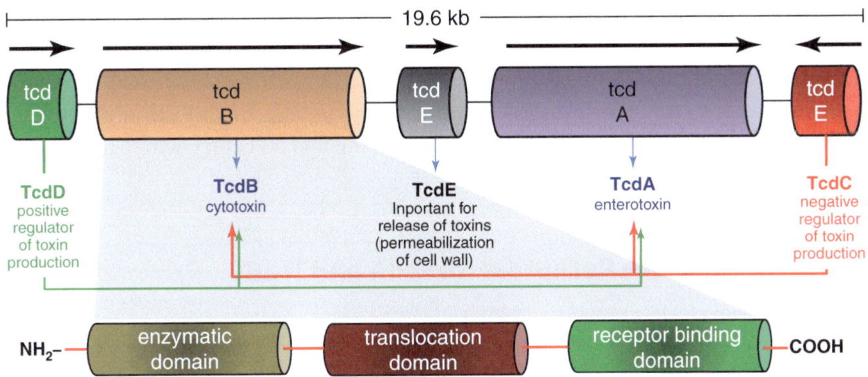


**Fig. 3.1**  Genetic arrangement of the *C. difficile* pathogenicity locus and proposed domain structures of TcdA and TcdB

## 3.4    Toxinotypes of *C. difficile* Isolates

In clinical isolates several genetic variants of TcdA and TcdB have been found. At least 22 toxinotypes of *C. difficile* have been described [16]. If toxinotyping will be useful in predicting the degree of virulence of particular isolates is not clear. It remains to be seen if it will be possible to anticipate the severity of disease based on the toxinotype found in infected patients.

## 3.5    Environmental Signals Regulating Toxin Expression

The precise environmental signals modulating toxin expression remain unclear, but in vitro studies indicate that toxin expression may be enhanced by stress (including antibiotics) and catabolite repression [17]. Given the role of antibiotic therapy in initiating pseudomembranous colitis, early studies focused on the influence of various antimicrobials on toxin production and found that subinhibitory levels of penicillin and vancomycin enhanced toxin production in continuous cultures of *C. difficile* [18].

Limiting biotin in defined medium enhances toxin production [19], and this phenomenon is striking, with a 64-fold increase in TcdB production and a 35-fold increase in TcdA when *C. difficile* is grown in the presence of 0.05 nM biotin. Under these biotin-limiting conditions, growth of *C. difficile* was reduced, and it is possible that this stress, like other stresses, triggered toxin production. In contrast, limiting amino acids does not enhance toxin production [20].

## 3.6    Mechanism of Action and Functional Domains
##         of TcdA and TcdB

TcdA and TcdB utilize a well-defined mechanism of action in order to modulate cell physiology and to alter the host environment. Both toxins target the Ras superfamily of small GTPases for modification via glycosylation. This irreversible modification inactivates these small regulatory proteins, leading to disruption of vital signaling pathways in the cell. However, in addition to enzymatic modification of targets, there are several important steps in receptor binding and cell entry which are important for intoxication.

## 3.7    Cell Surface Receptors for TcdA and TcdB

In order to elicit cytotoxic effects, TcdA and TcdB must be internalized into the host cell via endocytosis and require an acidic endosome for access to the cytosol [21–23]. Receptor binding is the first essential step in the process of cell entry. The receptor for the large clostridial toxins is thought to be non-proteinaceous and, for TcdA, show the disaccharide Galβ1-4GlcNac. The disaccharide is found in I, X, and

Y blood antigens present on a variety of cells, and these antigens have been shown to act as receptors for TcdA [24]. In line with this is the fact that treatment of cells or tissue with galactosidase reduces the binding of TcdA [25, 26]. TcdB has been shown to intoxicate a broad range of cell types, indicating that the receptor for this toxin, while undefined, is ubiquitous [1].

## 3.8   Receptor-Mediated Endocytosis and Membrane Translocation

Both TcdA and TcdB enter the host cell through receptor-mediated endocytosis and require an acidified endosome for translocation. The cytotoxic effect can be directly inhibited by a variety of lysosomotropic inhibitors, such as bafilomycin A and ammonium chloride [21]. Furthermore, conditions that prevent lysosome fusion with the endosome attenuate TcdB activity [27]. The requirement for low pH appears to be due to important structural changes which occur in the toxins, leading to exposure of hydrophobic domains prior to insertion into the target membrane [28].

## 3.9   Modification of Small GTPase Proteins as Host Cell Targets of TcdA and TcdB

Following receptor binding and internalization into the target cell cytosol, TcdA and TcdB are acting on small GTPase proteins of the host cell, which finally results in the destructions of the cell's actin cytoskeleton [23, 29, 30].

The small GTPase proteins, Rho, Rac, and Cdc42, are important regulators of the host cell's cytoskeleton. Rho, which includes mammalian isoforms A, B, and C, is a primary regulator of the actin cytoskeleton and is found ubiquitously in eukaryotic cells [31]. Following expression, Rho is subject to posttranslational modifications by prenylation and carboxymethylation, which aid in localization to the cytoplasmic side of the plasma membrane, where effective exchange of GTP and GDP occurs [32, 33]. Small GTPases bind and hydrolyze GTP to GDP, and it is the alteration between the GTP-bound (active) and GDP-bound (inactive) states that regulates activity (Fig. 3.2).

The exchange of GDP for GTP is regulated by guanine exchange factors within the cell [34]. Activated Rho intrinsically hydrolyzes GTP, removing the $\gamma$-phosphate group and returning to the "off" GDP-bound state through a process enhanced by GTPase-activating proteins [35, 36]. Further regulations come from GDP dissociation inhibitors, which block nucleotide exchange and maintain GTPases in the GDP-bound form within the cytoplasm, preventing association of Rho with the membrane and interaction with guanine exchange factors [37–39].

Rho plays a major role in regulation of stress fiber formation within the cell, and this is accomplished by interacting with and activating signaling proteins such as Rho-kinase [40], citron K [41], and phosphatidylinositol 4-phosphate 5-kinase [42], following activation from extracellular signals and integrin binding (Fig. 3.3).

**Fig. 3.2** Mechanism of action of the GTPase Rho

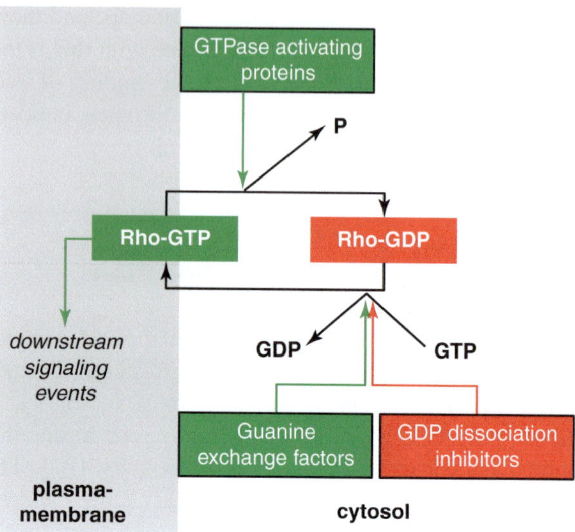


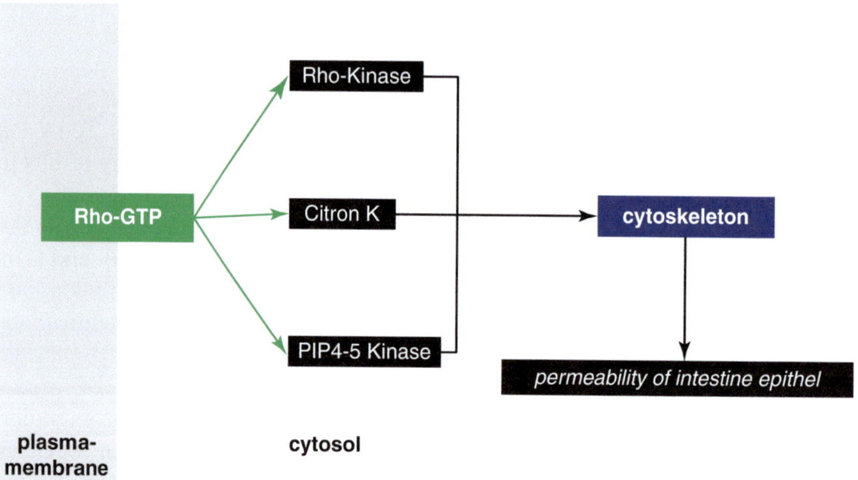


**Fig. 3.3**  Downstream effects regulated by activated Rho

Inhibiting Rho's ability to interact with these effectors leads to several changes in the actin cytoskeleton. Important to the role of TcdA and TcdB in disease is the fact that Rho regulates the localization of stress fibers at focal adhesion sites [43]. Furthermore, Rho controls the formation of perijunctional rings at the apical side of epithelial cells [44]. Inactivation of Rho and subsequent loss of perijunctional rings and focal adhesions sites may lead to increased permeability of the epithelial layer of the intestine.

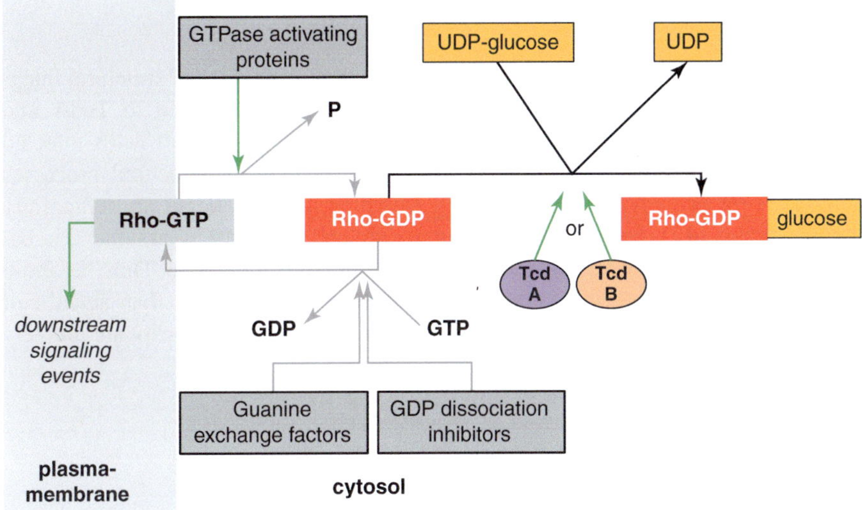


**Fig. 3.4** Inhibition of GTPase Rho by TcdA or TcdB

Due to their glucosyltransferase activities, both TcdA and TcdB can glycosylate Rho (and other small GTPases) by using UDP-glucose as a co-substrate [45]. This modification results in trapping Rho in its inactive GDP-bound state (Fig. 3.4).

The "inactivation" of Rho prevents all downstream events which results in profound changes of the host cell actin cytoskeleton. In a first phase actin condensates and the host cell starts to round. This is followed by a process called membrane blebbing. Blebbing is part of the process of apoptosis (programmed cell death) during which the cell's cytoskeleton breaks up and causes the cell membrane to bulge outward. These bulges may separate from the cell, taking a portion of cytoplasm with them, to become known as apoptotic blebs. The final result of this process is the death of the target cell.

While Rho regulates stress fiber formation, motility, and focal adhesion, Rac and Cdc42 are more specifically involved with lamellipodium and filopodium formation [46]. Lamellipodia and filopodia are important structural extensions at the leading edge of migrating cells and are essential to movement and sensing the environment during cell migration. Rac and Cdc42 show similar mechanisms of regulations as has been described for Rho.

## 3.10   Influence of TcdA and TcdB on Cell Physiology

TcdA and TcdB have a dramatic influence on mammalian cell physiology, altering events ranging from cell signaling to ultrastructure maintenance. In many cases cells do not survive intoxication by TcdA and TcdB, even at relatively low doses in comparison with other cytotoxins. Thus, cell death is one of the most obvious impacts of these toxins on cell physiology.

## 3.11    Impact on Cell Morphology

The first and most obvious impact on cell physiology is the loss of structural integrity, which results from the decline in F-actin in cells exposed to TcdA and TcdB. Mechanistically, the loss of structural integrity in TcdA- and TcdB-infected cells is expected, since Rho, Rac, and Cdc42 each regulate structural processes dependent on actin polymerization. Cell rounding and cell death are temporally distinct events [47]. TcdB is capable of inducing cell rounding in less than 2 h, but cell death does not occur until almost 24 h following treatment. Thus, as these events relate to disease, cytopathic effects may be more relevant than actual cell death, since cells with considerable loss of actin cytoskeleton integrity are unlikely to carry out their roles within the host.

## 3.12    Impact on Small GTPase Signaling Pathways

The substrate targets of TcdA and TcdB, Rho, Rac, and Cdc42, also regulate intracellular events beyond the actin cytoskeleton (Fig. 3.5). For example, Rho, Rac, and Cdc42 are involved in the regulation of the cell cycle and signaling through mitogen-activated protein kinases. The temporal differences between cell rounding and cell death indicate that other events downstream of actin condensation can contribute to cell death.

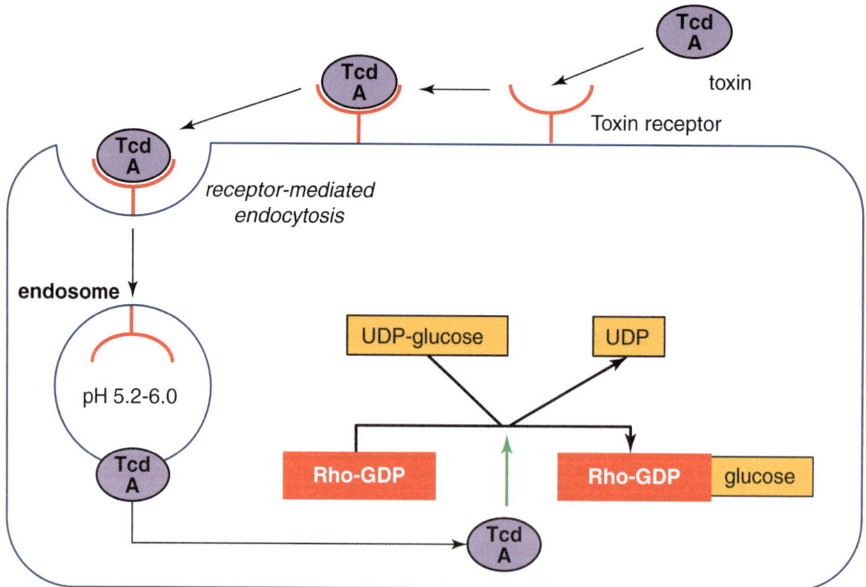


**Fig. 3.5**  Overview of intracellular actions of TcdA or TcdB

TcdA is capable of increasing permeability of colonic epithelial layers by inactivating Rho A through a process involving protein kinase C [48]. This increase in permeability may further enhance the inflammatory events observed in pseudomembranous colitis by increasing access of polymorphonuclear cells to the colonic epithelium. TcdA activates differential expression of chemokines in human intestinal epithelial cells, and this may also serve as a mechanism to increase the inflammatory response [49]. Furthermore, TcdA has been shown to activate mitogen-activated protein kinases in human THP-1 monocytes, in a process that appears to be independent of Rho inactivation and linked to interleukin-8 [50].

Apoptosis likely accounts for the mechanism of death in cells exposed to TcdA and TcdB. Rho inactivation in an endothelial cell line resulted in activation of caspase-3 and caspase-9, key components of the apoptotic cell death pathway [51]. There are temporal and spatial reasons to suggest that TcdA can induce apoptosis outside of inactivating Rho proteins. Treatment of cells with TcdA results in accumulation of the toxin at the mitochondria within 5 min following exposure, and this localization event occurs before detectable glucosylation of Rho proteins [52]. Thus, TcdA may induce apoptosis by disrupting mitochondria, which promotes pro-apoptotic events.

TcdB is also capable of triggering apoptosis by caspase-dependent and caspase-independent processes. The proteome of TcdB-treated cells contains fragments of intermediate filaments that result from cleavage by caspase-3. Since intermediate filaments, such as vimentin, are involved in maintaining the cell's structural integrity, the loss of these proteins may also account for changes in cell morphology. Together, these findings demonstrate some of the potential effects of GTPase inactivation on the host cell due to intoxication and implicate that apoptosis is a major process by which the toxins cause cell death.

## 3.13   Role of TcdA and TcdB in Disease

The signs of *C. difficile* disease range from mild diarrhea to fulminant colitis in patients undergoing antibiotic treatment. The disease occurs almost exclusively in the large bowel and shows distinguishing microscopic and gross lesions. Microscopic pathology reveals the "volcanic eruption" characteristics of the pseudomembranous lesion observed in pseudomembranous colitis, although this is not a defining pathology of all *C. difficile*-associated disease [53]. Gross pathology shows hallmark raised plaques within the intestines, which correspond to the nodules observed by endoscopy. These pathologies may explain the malabsorption and resulting diarrhea in patients with pseudomembranous colitis. Most of these pathologies can be ascribed to inflammatory events, which are likely modulated by TcdA and TcdB.

Both TcdA and TcdB are capable of mimicking the physiological events occurring in *C. difficile*-related pseudomembranous colitis [4, 5]. As shown in Fig. 3.6, inflammation, including increased epithelial permeability [54], cytokine and chemokine production [55, 56],

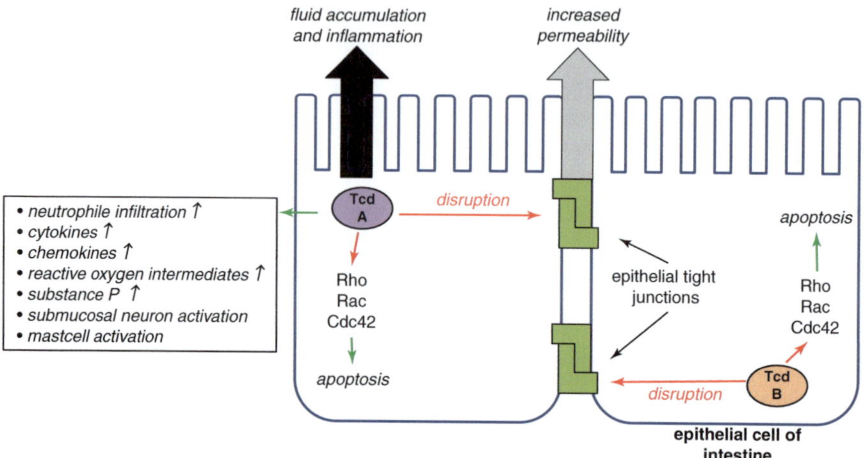



**Fig. 3.6**  Overview of extracellular actions of TcdA or TcdB

neutrophil infiltration [57, 58], activation of submucosal neurons [59], production of reactive oxygen intermediates [52], mast cell activation [60], substance P production [61], and direct damage to the intestinal mucosa contribute to the pathologies observed in pseudomembranous colitis.

One of the most direct events attributed to TcdA and TcdB during pseudomembranous colitis is the toxins' ability to disrupt tight junctions of the epithelial barrier [54]. This is likely due to inactivation of Rho proteins, since these small GTPases are known to be important in maintaining tight junctions [44]. In addition to directly altering the integrity of the epithelium, loss of tight junctions may provide a convenient conduit for migration of neutrophils into the intestine. Neutrophil accumulation is a hallmark of pseudomembranous colitis, and these cells contribute significantly to pseudomembrane formation [62]. Recruitment of neutrophils to the site of infection during pseudomembranous colitis is proposed to occur in response to direct interaction with TcdA and as a result of secondary events mediated by inflammatory molecules [57]. A direct binding of TcdA to neutrophils has been demonstrated [58].

## 3.14  Other Toxins Produced by *C. difficile*

In addition to TcdA and TcdB, a limited number of isolates have also been found to produce a third toxin named CDT. This noncytotoxic toxin is a two-component ADP-ribosyltransferase toxin encoded by the genes cdtA and cdtB [63]. The dimeric toxin formed by the proteins CDTa and CDTb are capable of modifying actin by its ADP-ribosyltransferase activity [64]. Examination of a larger number of *C. difficile* isolates showed that the binary toxin is found in about 15% of all isolates [65]. Strains encoding TcdA and TcdB as well as CDT do not appear to be more virulent,

**Table 3.1** Frequency distribution of *C. difficile* toxin profiles by source

| | Number (%) of patients | | |
| | Symptomatic patients | | Asymptomatic patients |
| Toxin profile | HA-CDI ($n = 96$) | CA-CDI ($n = 152$) | TCD ($n = 76$) |
| --- | --- | --- | --- |
| A$^+$, B$^+$, CDT$^+$ | 4 (4.2) | 7 (4.6) | 3 (4.0) |
| A$^+$, B$^+$, CDT$^-$ | 83 (86.5) | 139 (91.4) | 71 (93.4) |
| A$^-$, B$^+$, CDT$^+$ | 1 (1.0) | 2 (1.3) | 0 (0.0) |
| A$^-$, B$^+$, CDT$^-$ | 1 (1.0) | 1 (0.7) | 1 (1.3) |
| A$^-$, B$^-$, CDT$^+$ | 0 (0.0) | 0 (0.0) | 1 (1.3) |
| A$^-$, B$^-$, CDT$^-$ Nontoxigenic | 7 | 3 | |

*HA* hospital acquired, *CA* community acquired, *CDI C. difficile* infection, *TCD* toxigenic *C. difficile* colonization

nor does the binary toxin seem to have any notable effects in animal models. Thus, while the roles of TcdA and TcdB are well established and essential to *C. difficile* virulence, the possible contribution of CDT is not well understood.

## 3.15 Toxin Profiles of Clinical Isolates

Five different toxin profiles have been identified among toxigenic *C. difficile* strains isolated from patients (Table 3.1) [66]. The proportion of toxin profiles did not significantly differ between *C. difficile* categories. The most common toxin profile was toxin A-positive, toxin B-positive, and CDT-negative (83.2%). Toxin A-negative, toxin B-positive, and CDT-positive *C. difficile* isolates were recovered only from symptomatic patients, while a toxin A- and toxin B-negative and CDT-positive isolate was recovered from only one asymptomatic patient. Nontoxigenic *C. difficile* strains were isolated from 10 symptomatic patients (7 with HA-CDI, 3 with CA-CDI), most likely due to coinfection with TCD strains that were not isolated.

## References


1. Voth DE, Ballard JB. Clostridium difficile toxins: mechanism of action and role in disease. Clin Microbiol Rev. 2005;18(2):247–63. https://doi.org/10.1128/CMR.18.2.247-263.2005.
2. Burdon DW, George RH, Mogg GA, Arabi Y, Thompson H, Johnson M, Alexander-Williams J, Keighley MR. Faecal toxin and severity of antibiotic-associated pseudomembranous colitis. J Clin Pathol. 1981;34(5):548–51. https://doi.org/10.1136/jcp.34.5.548.
3. Kyne L, Warny M, Qamar A, Kelly CP. Association between antibody response to toxin A and protection against recurrent Clostridium difficile diarrhea. Lancet. 2001;357(9251):189–93. https://doi.org/10.1016/S0140-6736(00)03592-3.
4. Lima AA, Lyerly DM, Wilkins TD, Innes DJ, Guerrant RL. Effects of Clostridium difficile toxins A and B in rabbit small and large intestine in vivo and on cultured cells in vitro. Infect Immun. 1988;56(3):582–8.

5. Savidge TC, Pan WH, Newman P, O'Brien M, Anton PM, Pothoulakis C. Clostridium difficile toxin B is an inflammatory enterotoxin in human intestine. Gastroenterology. 2003;125(2):413–20. https://doi.org/10.1016/S0016-5085(03)00902-8.

6. Lyerly DM, Saum KE, MacDonald DK, Wilkins TD. Effects of Clostridium Difficile toxins given intragastrically to animals. Infect Immun. 1985;47(2):349–52.

7. Alfa MJ, Kabani A, Lyerly D, Moncrief S, Neville LM, Al-Barrak A, Harding GKH, Dyck B, Olekson K, Embil JM. Characterization of a toxin A-negative, toxin B-positive strain of Clostridium difficile responsible for a nosocomial outbreak of Clostridium difficile-associated diarrhea. J Clin Microbiol. 2000;38(7):2706–14.

8. Barroso LA, Wang SZ, Phelps CJ, Johnson JL, Wilkins TD. Nucleotide sequence of Clostridium difficile toxin B gene. Nucleic Acids Res. 1990;18(13):4004. https://doi.org/10.1093/nar/18.13.4004.

9. Dove CH, Wang SZ, Price SB, Phelps CJ, Lyerly DM, Wilkins TD, Johnson JL. Molecular characterization of the Clostridium difficile toxin A gene. Infect Immun. 1990;58(2):480–8.

10. von Eichel-Streiber C, Laufenberg-Feldmann R, Sartingen S, Schulze J, Sauerborn M. Comparative sequence analysis of the Clostridium difficile toxins A and B. Mol Gen Genet. 1992;233(1–2):260–8. https://doi.org/10.1007/bf00587587.

11. von Eichel-Streiber C, Laufenberg-Feldmann R, Sartingen S, Schulze J, Sauerborn M. Cloning of Clostridium difficile toxin B gene and demonstration of high N-terminal homology between toxin A and B. Med Microbiol Immunol. 1990;179:271–9. https://doi.org/10.1007/BF00192465.

12. Hammond GA, Johnson JL. The toxigenic element of Clostridium difficile strain VPI 10463. Microbial Pathogenesis. 1995;19(4):203–13. https://doi.org/10.1016/S0882-4010(95)90263-5.

13. Hundsberger T, Braun V, Weidmann M, Leukel P, Sauerborn M, von Eichel-Streiber C. Transcription analysis of the genes tcdA-E of the pathogenicity locus of Clostridium difficile. Eur J Biochem. 1997;244(3):735–42. https://doi.org/10.1111/j.1432-1033.1997.t01-1-00735.x.

14. Moncrief JS, Barroso LA, Wilkins TD. Positive regulation of Clostridium difficile toxins. Infect Immun. 1997;65(3):1105–8.

15. Tan KS, Wee BY, Song KP. Evidence for holin function of tcdE gene in the pathogenicity of Clostridium difficile. J Med Microbiol. 2001;50(7):613–9. https://doi.org/10.1099/0022-1317-50-7-613.

16. Geric B, Rupnik M, Gerding DN, Grabnar M, Johnson S. Distribution of Clostridium difficile variant toxinotypes and strains with binary toxin genes among clinical isolates in an American hospital. J Med Microbiol. 2004;53(9):887–94. https://doi.org/10.1099/jmm.0.45610-0.

17. Dupuy B, Sonenshein AL. Regulated transcription of Clostridium difficile toxin genes. Mol Microbiol. 1998;27(1):107–20. https://doi.org/10.1046/j.1365-2958.1998.00663.x.

18. Onderdonk AB, Lowe BR, Bartlett JG. Effect of environmental stress on Clostridium difficile toxin levels during continuous cultivation. Appl Environ Microbiol. 1979;38(4):637–41.

19. Yamakawa K, Karasawa T, Ikoma S, Nakamura S. Enhancement of Clostridium difficile toxin production in biotin-limited conditions. J Med Microbiol. 1996;44(2):111–4.

20. Yamakawa K, Kamiya S, Meng XQ, Karasawa T, Nakamura S. Toxin production by Clostridium difficile in a defined medium with limited amino acids. J Med Microbiol. 1994;41(5):319–23. https://doi.org/10.1099/00222615-41-5-319.

21. Florin I, Thelestam M. Internalization of Clostridium difficile cytotoxin into cultured human lung fibroblasts. Biochim Biophys Acta. 1983;763(4):383–92. https://doi.org/10.1016/0167-4889(83)90100-3.

22. Henriques B, Florin I, Thelestam M. Cellular internalisation of Clostridium difficile toxin A. Microb Pathog. 1987;2(6):455–63. https://doi.org/10.1016/0882-4010(87)90052-0.

23. Mitchell MJ, Laughon BE, Lin S. Biochemical studies on the effect of Clostridium difficile toxin B on actin in vivo and in vitro. Infect Immun. 1987;55(7):1610–5.

24. Tucker KD, Wilkins TD. Toxin A of Clostridium difficile binds to the human carbohydrate antigens I, X, and Y. Infect Immun. 1991;59(1):73–8.

25. Krivan HC, Clark GF, Smith DF, Wilkins TD. Cell surface binding site for Clostridium difficile enterotoxin: evidence for a glycoconjugate containing the sequence gal alpha 1-3Gal beta 1-4GlcNAc. Infect Immun. 1986;53(3):573–81.

26. Smith JA, Cooke DL, Hyde S, Borriello SP, Long RG. Clostridium difficile toxin A binding to human intestinal epithelial cells. J Med Microbiol. 1997;46:953–8. https://doi.org/10.1099/00222615-46-11-953.

27. Florin I, Thelestam M. Lysosomal involvement in cellular intoxication with Clostridium difficile toxin B. Microb Pathog. 1986;1(4):373–85. https://doi.org/10.1016/0882-4010(86)90069-0.

28. Qa'Dan M, Spyres LM, Ballard JB. pH-induced conformational changes in Clostridium difficile toxin B. Infect Immun. 2000;68(5):2470–4. https://doi.org/10.1128/iai.68.5.2470-2474.2000.

29. Thelestam M, Brönnegård M. Interaction of cytopathogenic toxin from Clostridium difficile with cells in tissue culture. Scand J Infect Dis Suppl. 1980;(Suppl 22):16–29.

30. Wedel N, Toselli P, Pothoulakis C, Faris B, Oliver P, Franzblau C, LaMont T. Ultrastructural effects of Clostridium difficile toxin B on smooth muscle cells and fibroblasts. Exp Cell Res. 1983;148(2):413–22. https://doi.org/10.1016/0014-4827(83)90163-5.

31. Hall A. The cellular functions of small GTP-binding proteins. Science. 1990;249(4969):635–40. https://doi.org/10.1126/science.2116664.

32. Adamson P, Marshall CJ, Hall A, Tilbrook PA. Post-translational modifications of p21rho proteins. J Biol Chem. 1992;267:20033–8.

33. Marshall CJ. Protein prenylation: a mediator of protein-protein interactions. Science. 1993;259(5103):1865–6. https://doi.org/10.1126/science.

34. Zhou K, Wang Y, Gorski JL, Nomura N, Collard J, Bokoch GM. Guanine nucleotide exchange factors regulate specificity of downstream signaling from Rac and Cdc42. J Biol Chem. 1998;273:16782–6. https://doi.org/10.1074/jbc.273.27.16782.

35. Garrett MD, Self AJ, van Oers C, Hall A. Identification of distinct cytoplasmic targets for ras/R-ras and rho regulatory proteins. J Biol Chem. 1989;264:10–3.

36. Garrett MD, Major GN, Totty N, Hall A. Purification and N-terminal sequence of the p21rho GTPase-activating protein, rho GAP. Biochem J. 1991;276(Pt 3):833–6. https://doi.org/10.1042/bj2760833.

37. Fukumoto Y, Kaibuchi K, Hori Y, et al. Molecular cloning and characterization of a novel type of regulatory protein (GDI) for the rho proteins, ras p21-like small GTP-binding proteins. Oncogene. 1990;5(9):1321–8.

38. Hiraoka K, Kaibuchi K, Ando S, et al. Both stimulatory and inhibitory GDP/GTP exchange proteins, smg GDS and rho GDI, are active on multiple small GTP-binding proteins. Biochem Biophys Res Commun. 1992;182(2):921–30. https://doi.org/10.1016/0006-291x(92)91820-g.

39. Mizuno T, Kaibuchi K, Yamamoto T, Kawamura M, Sakoda T, Fujioka H, Matsuura Y, Takai Y. A stimulatory GDP/GTP exchange protein for smg p21 is active on the post-translationally processed form of c-Ki-ras p21 and rhoA p21. Proc Natl Acad Sci U S A. 1991;88(15):6442–6. https://doi.org/10.1073/pnas.88.15.6442.

40. Fujisawa K, Madaule P, Ishizaki T, Watanabe G, Bito H, Saito Y, Hall A, Narumiya S. Different regions of rho determine rho-selective binding of different classes of rho target molecules. J Biol Chem. 1998;273:18943–9. https://doi.org/10.1074/jbc.273.30.18943.

41. Madaule P, Eda M, Watanabe N, Fujisawa K, Matsuoka T, Bito H, Ishizaki T, Narumiya S. Affiliations expand et al. Role of citron kinase as a target of the small GTPase Rho in cytokinesis. Nature. 1998;394(6692):491–4. https://doi.org/10.1038/28873.

42. Chong LD, Traynor-Kaplan A, Bokoch GM, Schwartz MA. The small GTP-binding protein rho regulates a phosphatidylinositol 4-phosphate 5-kinase in mammalian cells. Cell. 1994;79(3):507–13.

43. Ridley AJ, Hall A. The small GTP-binding protein rho regulates the assembly of focal adhesions and actin stress fibers in response to growth factors. Cell. 1992;70(3):389–99. https://doi.org/10.1016/0092-8674(92)90163-7.

44. Nusrat A, Giry M, Turner JR, Colgan SP, Parkos CA, Carnes D, Lemichez E, Boquet P, Madara JL. Rho protein regulates tight junctions and perijunctional actin organization in polar-

ized epithelia. Proc Natl Acad Sci U S A. 1995;92(23):10629–33. https://doi.org/10.1073/pnas.92.23.10629.

45. Just I, Selzer J, Wilm M, von Eichel-Streiber C, Mann M, Aktories K. Glucosylation of Rho proteins by Clostridium difficile toxin B. Nature. 1995;375:500–3. https://doi.org/10.1038/375500a0.

46. Nobes CD, Hall A. Rho, rac, and cdc42 GTPases regulate the assembly of multimolecular focal complexes associated with actin stress fibers, lamellipodia, and filopodia. Cell. 1995;81(1):53–62. https://doi.org/10.1016/0092-8674(95)90370-4.

47. Qa'Dan M, Ramsey M, Daniel J, Spyres LM, Safiejko-Mroczka B, Ortiz-Leduc W, Ballard JD. Clostridium difficile toxin B activates dual caspase-dependent and caspase-independent apoptosis in intoxicated cells. Cell Microbiol. 2002;4:425–34. https://doi.org/10.1046/j.1462-5822.2002.00201.x.

48. Chen ML, Pothoulakis C, LaMont JT. Protein kinase C signaling regulates ZO-1 translocation and increased paracellular flux of T84 colonocytes exposed to Clostridium difficile toxin A. J Biol Chem. 2002;277:4247–54. https://doi.org/10.1074/jbc.M109254200.

49. Kim JM, Kim JS, Jung HC, Oh Y, Song IS, Kim CY. Differential expression and polarized secretion of CXC and CC chemokines by human intestinal epithelial cancer cell lines in response to Clostridium difficile toxin A. Microbiol Immunol. 2002;46:333–42. https://doi.org/10.1111/j.1348-0421.2002.tb02704.x.

50. Warny M, Keates AC, Castagliuolo I, Zacks JK, Aboudola S, Qamar A, Pothoulakis C, LaMont JL, Kelly CP. p38 MAP kinase activation by Clostridium difficile toxin A mediates monocyte necrosis, IL-8 production, and enteritis. J Clin Invest. 2000;105(8):1147–56. https://doi.org/10.1172/JCI7545.

51. Hippenstiel S, Schmeck B, N'Guessan PD, Seybold J, Krüll M, Preissner K, Eichel-Streiber CV, Suttorp N. Rho protein inactivation induced apoptosis of cultured human endothelial cells. Am J Physiol Lung Cell Mol Physiol. 2002;283(4):L830–8. https://doi.org/10.1152/ajplung.00467.

52. He D, Hagen SJ, Pothoulakis C, Chen M, Medina ND, Warny M, LaMont JT. Clostridium difficile toxin A causes early damage to mitochondria in cultured cells. Gastroenterology. 2000;119(1):139–50. https://doi.org/10.1053/gast.2000.8526.

53. Hurley BW, Nguyen CC. The spectrum of pseudomembranous enterocolitis and antibiotic-associated diarrhea. Arch Intern Med. 2002;162(19):2177–84. https://doi.org/10.1001/archinte.162.19.2177.

54. Feltis BA, Wiesner SM, Kim AS, Erlandsen SL, Lyerly DL, Wilkins TD, Wells CL. Clostridium difficile toxins A and B can alter epithelial permeability and promote bacterial paracellular migration through HT-29 enterocytes. Shock (Augusta, Ga.). 2000;14(6):629–34. https://doi.org/10.1097/00024382-200014060.

55. Castagliuolo I, Keates AC, Wang CC, Pasha A, Valenick L, Kelly CP, Nikulasson ST, JT LM, Pothoulakis C. Clostridium difficile toxin A stimulates macrophage-inflammatory protein-2 production in rat intestinal epithelial cells. J Immunol. 1998;160(12):6039–45.

56. He D, Sougioultzis S, Hagen S, Liu J, Keates S, Keates AC, Pothoulakis C, LaMont JT. Clostridium difficile toxin A triggers human colonocyte IL-8 release via mitochondrial oxygen radical generation. Gastroenterology. 2002;122(4):1048–57.

57. Kelly CP, Pothoulakis C, LaMont JT. Clostridium difficile colitis. N Engl J Med. 1994a;330:257. https://doi.org/10.1056/NEJM199401273300406.

58. Kelly CP, Becker S, Linevsky JK, Joshi MA, O'Keane JC, Dickey BF, LaMont JT, Pothoulakis C. Neutrophil recruitment in Clostridium difficile toxin A enteritis in the rabbit. J Clin Invest. 1994b;93(3):1257–65. https://doi.org/10.1172/JCI117080.

59. Neunlist M, Barouk J, Michel K, Just I, Oreshkova T, Schemann M, Galmiche JP. Toxin B of Clostridium Difficile activates human VIP submucosal neurons, in part via an IL-1beta-dependent pathway. Am J Physiol Gastrointest Liver Physiol. 2003;285(5):G1049–55. https://doi.org/10.1152/ajpgi.00487.2002.

60. Wershil BK, Castagliuolo I, Pothoulakis C. Direct evidence of mast cell involvement in Clostridium Difficile toxin A-induced enteritis in mice. Gastroenterology. 1998;114(5):956–64. https://doi.org/10.1016/s0016-5085(98)70315-4.

61. Mantyh CR, Pappas TN, Lapp JA, Washington MK, Neville LM, Ghilardi JR, Rogers SD, Mantyh PW, Vigna SR. Substance P activation of enteric neurons in response to intraluminal Clostridium difficile toxin A in the rat ileum. Gastroenterology. 1996;111(5):1272–80. https://doi.org/10.1053/gast.1996.v111.pm8898641.

62. Souza MH, Melo-Filho AA, Rocha MF, Lyerly DM, Cunha FQ, Lima AA, Ribeiro RA. The involvement of macrophage-derived tumour necrosis factor and lipoxygenase products on the neutrophil recruitment induced by Clostridium difficile toxin B. Immunology. 1997;91(2):281–8. https://doi.org/10.1046/j.1365-2567.1997.00243.x.

63. Perelle S, Gibert M, Bourlioux P, Corthier G, Popoff MR. Production of a complete binary toxin (actin-specific ADP-ribosyltransferase) by Clostridium difficile CD196. Infect Immun. 1997;65(4):1402–7.

64. Popoff MR, Rubin EJ, Gill DM, Boquet P. Actin-specific ADP-ribosyltransferase produced by a Clostridium difficile strain. Infect Immun. 1988;56(9):2299–306.

65. Geric B, Johnson S, Gerding DN, Grabnar M, Rupnik M. Frequency of binary toxin genes among Clostridium difficile strains that do not produce large Clostridial toxins. J Clin Microbiol. 2003;41(11):5227–32. https://doi.org/10.1128/JCM.41.11.5227-5232.2003.

66. Furuya-Kanamori L, Riley TV, Paterson DL, Foster NF, Huber CA, Hong S, Harris-Brown T, Robson J, Clements AC. Comparison of Clostridium difficile ribotypes circulating in Australian hospitals and communities. J Clin Microbiol. 2016;55(1):216–25. https://doi.org/10.1128/JCM.01779-16.

## Contents



## 4.1   Historical Perspective

A clinical description of a CDI-like disease was possibly first reported at Johns Hopkins Hospital in 1892 [1]. A 22-year-old female underwent gastric surgery for a cicatrizing ulcer and developed mild diarrhea 10 days postoperatively. Her diarrhea progressed into frequent bloody stools and she expired 5 days later. Autopsy records indicated a "diphtheritic colitis" was observed in the small bowel. While this occurred in the pre-antibiotic era, the patient had received a boric acid stomach irrigation prior to surgery as a local antiseptic. In 1935, Hall and O'Toole first described *Bacillus difficilis*, a difficult-to-grow, obligate anaerobic, Gram-positive, spore-forming, cytotoxin-producing rod, isolated from the intestinal tract of healthy newborn infants [2]. Eventually it was renamed *C. difficile*, but because it had been isolated from healthy infants, there was no reason to believe it had any deleterious effects in humans. Even today, it is well recognized that infants can be colonized with *C. difficile*, which can persist during the first 2 years of life [3]. For decades *C. difficile* was not linked to pseudomembranous colitis (PMC) and little remained known about the organism, except that it was considered to be part of the normal intestinal ecology of infants. Notably, PMC, a hallmark of CDI, was rare before the widespread use of antibiotics

© The Author(s), under exclusive license to Springer Nature Switzerland AG 2021     35
H. Sommermeyer, J. Piątek, *Clostridioides difficile*,
https://doi.org/10.1007/978-3-030-81100-6_4

in the late 1940s and early 1950s [4]. Subsequently, reports of antibiotic-associated diarrhea and PMC became much more frequent and antibiotic-resistant *S. aureus* was implicated as the causative organism, based on routine stool culture [5]. In 1974, a strong relationship between clindamycin and PMC was identified, termed "clindamycin-colitis" [6]. A prospective study showed that of 200 consecutive clindamycin recipients, 21% developed diarrhea, which upon further endoscopic examination revealed that 50% had evidence of PMC [6]. Later studies demonstrated that administration of vancomycin to hamsters was protective against clindamycin-colitis, suggesting that certain gut bacteria play a role in the pathogenesis of PMC [7]. The concept at the time that broad-spectrum antibiotics were altering the indigenous gut microbiota and allowing for overgrowth of a pathogenic organism set the stage for our current understanding of CDI. In the late 1970s, an early notion that *S. aureus* was the major bacterial cause of antibiotic-associated PMC was fading and a series of investigations eventually led to *C. difficile* being reported as causative pathogen. Investigators initially reported a clostridial toxin was the cause of the cytopathic effect on the gut [8, 9]. In 1978, toxin-producing *C. difficile* was established as a cause of PMC [10]. Eventually, several investigators were able to isolate *C. difficile* from the stool of patients with PMC [11–13]. During the next decade, research established epidemiologic data, clinical features, diagnostic tests, and effective therapies. By the late 1980s many believed CDI was well understood and could be effectively managed with available diagnostic and therapeutic modalities [14]. Until the turn of the twenty-first century, CDI was largely viewed as a treatable complication of antimicrobial therapy and not viewed as a major public health threat. However, in the past 20 years *C. difficile* infection (CDI) has emerged as an increasingly prevalent and severe infectious disease worldwide.

## 4.2    Hospital- and Community-Acquired CDI

*C. difficile* has historically been considered a nosocomial pathogen associated with antibiotic exposure in hospitals. However, CDI epidemiology is rapidly changing and about one quarter of all infections are now originating from the community (Fig. 4.1) [15].

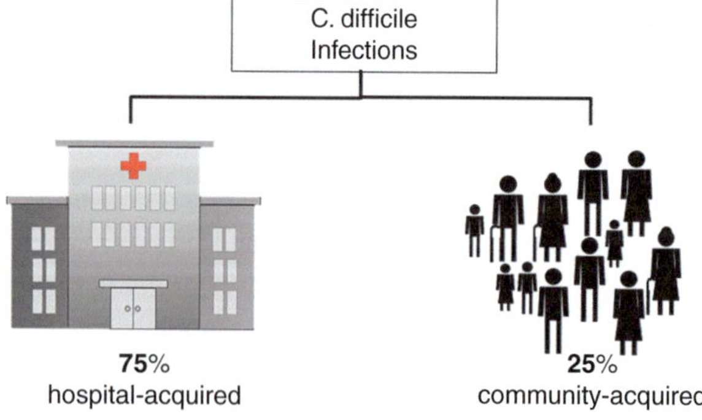


**Fig. 4.1** Ratio of hospital- and community-acquired CDI

**Table 4.1** Characteristics of patients with hospital-acquired (HA) and community-acquired (CA) *C. difficile* infection (CDI)

| Characteristics | CA-CDI | HA-CDI |
| --- | --- | --- |
| Median age [years] | 50 | 72 |
| Female | 72% | 60% |
| Comorbidity scores | Lower | Higher |
| Severe CDI | 20% | 31% |
| No antibiotics last 90 days | 22% | 6% |
| Gastric acid-suppressing therapy | 22% | 47% |
| Cancer | 17% | 32% |

An increased incidence is now observed in populations previously considered at low risk, such as healthy peripartum women, children, antibiotic-naive patients, and those with minimal or no recent healthcare exposure [16–19].

Data from North America and Europe suggest that approximately 20–27% of all CDI cases are community acquired, with an incidence of 20–30 per 100,000 of population [20]. In a population-based study an even higher rate (41%) was found [21]. The same study characterized patients with community-acquired CDI in comparison to patients with hospital-acquired CDI (Table 4.1).

The findings of the study demonstrate patients with community-acquired CDI are in general younger and less ill.

## 4.3   Hypervirulent *C. difficile* Strain Ribotype 027 (NAP1/BI)

Paralleling the increased CDI prevalence is an increase in morbidity and mortality, which has coincided with the emergence and rapid spread of a strain known synonymously as polymerase chain reaction (PCR) ribotype 027, North American Pulse-field type 1 (NAP1), or restriction endonuclease analysis (REA) type BI [22].

The story of *C. difficile* ribotype 027 begins in 2002 (Fig. 4.2), when an increased number of patients requiring colectomy alerted a hospital in Montreal, Quebec, Canada [23] to the possibility of CDI with a higher severity, mortality, and relapse rate. Over the next 2 years several investigations were performed: rates of CDI were 28/1000 admissions (five times the national average of 1997) with an extra 10.7 days in hospital and 30-day attributed mortality rates of 6.9%, compared to 0.8–2% in 1997 [24–27]. Perforations, toxic megacolon, and colectomy rates had also increased. Between 2004 and 2005 it was estimated that over 14,000 patients had been affected in the province of Quebec, with high mortality and relapse rates. By June 2006 the problem had spread to seven provinces with estimates of 13 cases per 1000 admissions [24, 25, 27]. Risk factors compared with matched controls [26] comprised, as in many previous studies, cephalosporins (odds ratio (OR) 3.8%, 95% confidence interval (CI) 2.2% to 6.6%) and, for the first time, fluoroquinolones (OR 3.9%, 95% CI 2.3% to 6.6%) [28].

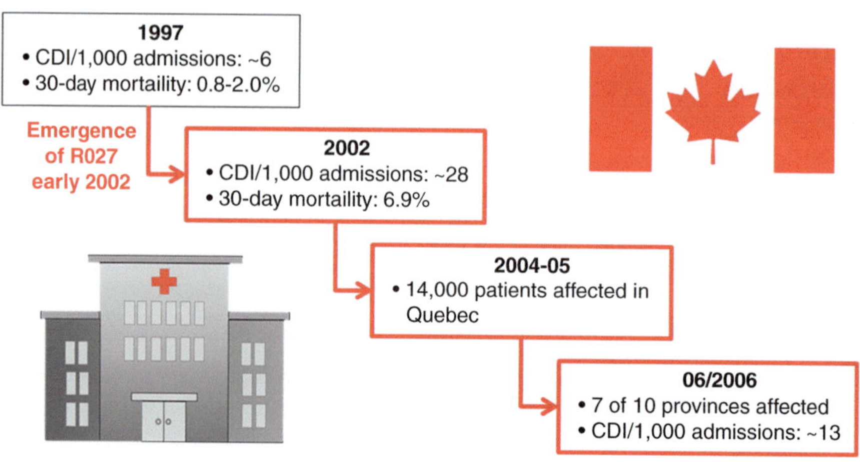

**Fig. 4.2** Spread of *C. difficile* ribotype 027 in Canada

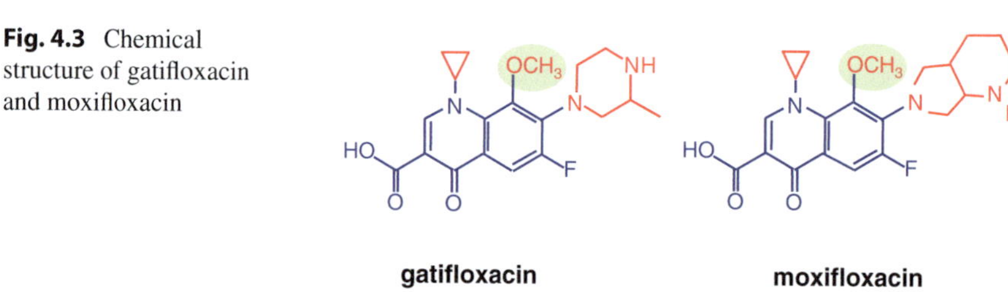

**Fig. 4.3** Chemical structure of gatifloxacin and moxifloxacin



In-depth analyses revealed that the risk association was only with quinolones of the 8-methoxy-quinolone type, like gatifloxacin and moxifloxacin (Fig. 4.3). These quinolones had only been introduced recently in several of the affected hospitals.

However, the situation with antibiotics was not straightforward, as patients received more antibiotics per case (46%) than controls [25, 26, 28]. *C. difficile* strain typing was performed and a new strain (ribotype 027) was isolated from 82% of patients. It was assumed that this was related to the increased virulence and relapse rate.

A similar situation to that in Canada emerged in the USA, with the Centers for Disease Control and Prevention (CDC) [29] showing rates had doubled from 31 to 61 cases per 100,000 between 1996 and 2003. Isolates from six outbreaks in the USA revealed emergence of the same epidemic ribotype 027 strain. Higher morbidity and mortality were described in at least 17 states [28, 30]. As of today, *C. difficile* ribotype 027 has spread around the world.

*C. difficile* ribotype 027 is a hypervirulent strain characterized by the hyperproduction of the toxins TcdA and TcdB. The reason for this high synthesis of TcdA and TcdB is a mutation in the tcdC gene in the pathogenicity locus of *C. difficile* (Fig. 4.4). The tcdD gene encodes the protein TcdC which is a negative regulator of the production of TcdA and TcdB. Due to the mutation TcdC cannot any longer be produced and the inhibition of toxin production is abolished.

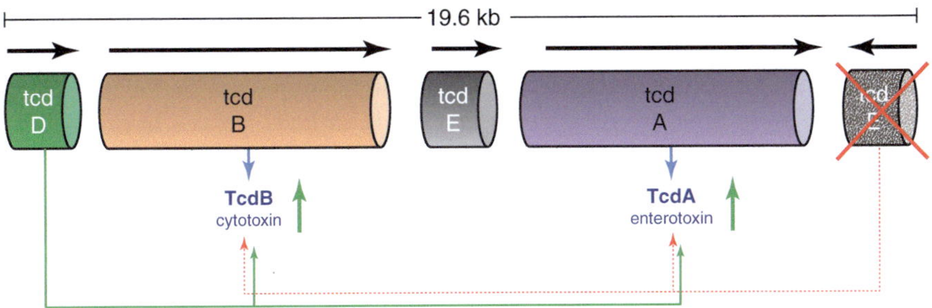

**Fig. 4.4** Mutation of tcdC in *C. difficile* ribotype 027 results in strongly increased production of TcdA and TcdB

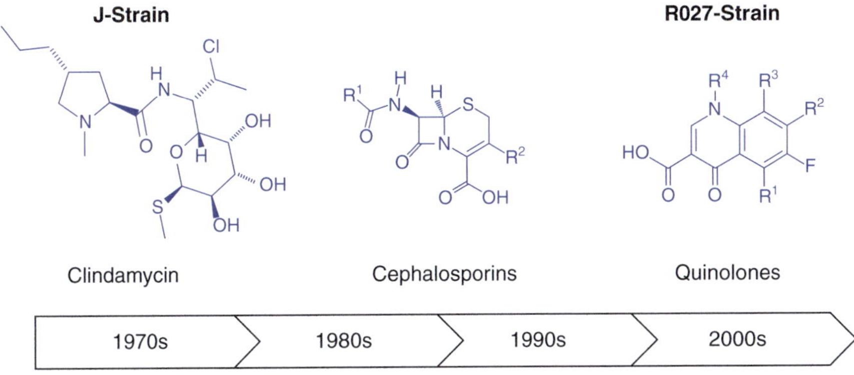


**Fig. 4.5** Resistance acquisitions of *C. difficile*

*C. difficile* ribotype 027 is also a producer of the *C. difficile* binary toxin CDTa/CDTb. In addition, mutations of surface layer proteins have resulted in an increased adherence of *C. difficile* ribotype 027 to the intestinal epithelium. The ribotype 027 strain is characterized by a higher sporulation rate compared to the wild type. *C. difficile* ribotype 027 is resistant to quinolones (mainly moxifloxacin and gatifloxacin), which makes this strain a recent development stage in *C. difficile* resistance acquisitions (Fig. 4.5).

Studies in North America and Europe showed that the ribotype 027 strain is characterized by (i) an increased incidence and severity, (ii) higher infection rates in older adults in hospitals or in long-term care facilities, (iii) refractory to traditional therapy, and (iv) a greater risk of relapse [25, 26, 30–33].

## 4.4   Other Strains of *C. difficile*

More recently, additional strains have been reported to cause CDI with increased severity and poor outcomes [34, 35]. Over recent years a large number of different ribotypes of *C. difficile* strains have been identified by the use of PCR methodology.

As part of the European, multicenter, prospective, biannual, point-prevalence study of *C. difficile* infection in hospitalized patients with diarrhea (EUCLID), the largest *C. difficile* epidemiological study of its type, PCR ribotype distribution of *C. difficile* isolates in Europe was determined in 1196 *C. difficile* isolates from diarrheal samples sent to the European coordinating laboratory in 2012/13 and 2013 (from two sampling days) by 482 participating hospitals from 19 European countries [36]. 125 distinct ribotypes were identified, with considerable intercountry variation in ribotype distribution. Ribotypes 027 (19%), 001/072 (11%), and 014/020 (10%) were the most prevalent, followed by ribotypes 002, 140, 010, 078, 176, and 018 (each <5%).

Analysis of *C. difficile* isolates from 19 countries demonstrated that 5 ribotypes had within-country clustering: ribotype 356, only in Italy; ribotype 018, predominantly in Italy; ribotype 176, with distinct Czech and German clades; ribotype 001/072, including distinct German, Slovakian, and Spanish clades; and ribotype 027, with multiple predominantly country-specific clades including in Hungary, Italy, Germany, Romania, and Poland. By contrast, no within-country clustering was observed for ribotypes 078, 015, 002, 014, and 020, which is consistent with a Europe-wide distribution [37]. These and other data support the existence of two distinct patterns of *C. difficile* ribotype spread, which are consistent with either predominantly healthcare-associated acquisition or Europe-wide dissemination via other routes/sources (for example, possibly via the food chain) [38, 39].

With enhanced DNA fingerprinting (multilocus variable repeat analysis, MLVA) it is possible to discriminate even within a particular ribotype. For example, MLVA can distinguish more than 20 sub-types of *C. difficile* ribotype 027 [40].

## 4.5    Dual Strain Infections

Evidence of CDI with more than one strain was first described in 1983 [41]. Infection with two different *C. difficile* strains is found in about 5–10% of CDI patients. Multiple strain infections can be detected by methods that can distinguish individual strains. These methods include restriction enzyme analysis (REA), pulsed field gel electrophoresis (PGFE), PCR ribotyping, multilocus variable number tandem repeat analysis (MLVA), multilocus sequence typing (MLST), and whole genome sequencing (WGS) [38, 42, 43].

The presence of multiple *C. difficile* strains could be the result of the ingestion of an additional strain when another has already colonized the host. It has been shown that colonization with a nontoxigenic strain of *C. difficile* protects against disease in hamster, following a challenge with a toxigenic strain [44]. Patients with an episode of primary CDI or first recurrence within 8 weeks of the primary episode benefitted from colonization with the nontoxigenic *C. difficile* strain M3 [45]. These findings indicate that the presence of two competing strains can be beneficial to the host as they could control each other, similar to how probiotics have been proposed for the prevention of CDI [46]. It has been observed that patients who are colonized by a nontoxigenic *C. difficile* strain are at lower risk of developing *C. difficile* infection while in hospital. This is supported by results from a Phase 2 randomized controlled trial using nontoxigenic *C. difficile* spores to prevent recurrent *C. difficile* infection [47].

## 4.6   CDI in Children

Asymptomatic colonization of *C. difficile* with either toxigenic or nontoxigenic strains is common in early infancy, and it often occurs in the first week of life [48–50]. Nontoxigenic strains are more common than toxigenic strains among colonized infants and different strains are found to colonize the same infant at different times [51–55]. The susceptibility to *C. difficile* colonization might be because of the immaturity of the intestine and lack of protective intestinal microbiota [56]. *C. difficile* carrier rates average 37% for infants 0–1 month of age and 30% between 1 month and 6 months of age [57]. The most likely source in infants is from environmental contamination rather than direct maternal infant transmission [53, 58, 59]. Breastfed infants have lower carriage rates than formula-fed infants (14% vs. 30%, respectively) [60]. At 6–12 months of age, approximately 14% of children are colonized with C difficile, and by 3 years of age, the rate is similar to that of non-hospitalized adults (0% to 3%) [57] (Fig. 4.6).

Clinical illness of CDI in infants is rarely reported before 12–24 months of age. The mechanisms for the resistance of infants to CDI are thought to be related to the immunoglobulin fractions of breast milk that inhibit the binding of toxin A to its intestinal receptor, and absence in the newborn gut of the intestinal receptor that binds *C. difficile* toxin A [56, 61, 62].

Similar to the findings in adults, the incidence of CDI has risen in children since 2000 [18, 63–65]. Carriage rates in hospitalized children approximate 20% [66]. Among the children hospitalized with CDI, 26% are younger than 1 year and 5% are

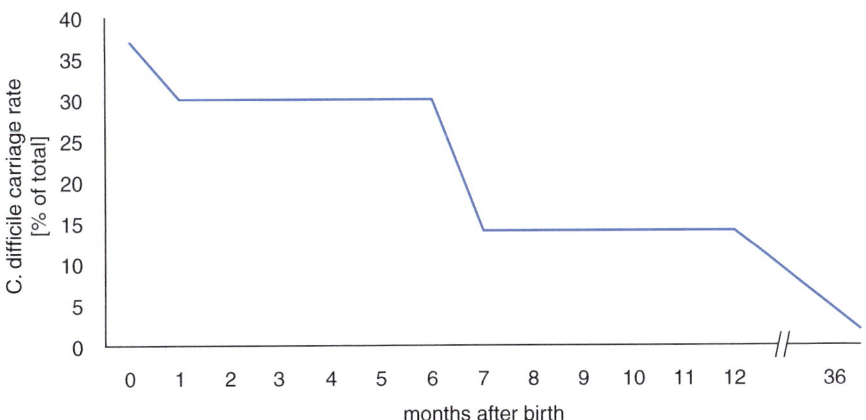


**Fig. 4.6**  *C. difficile* carrier rate in infants as function of age

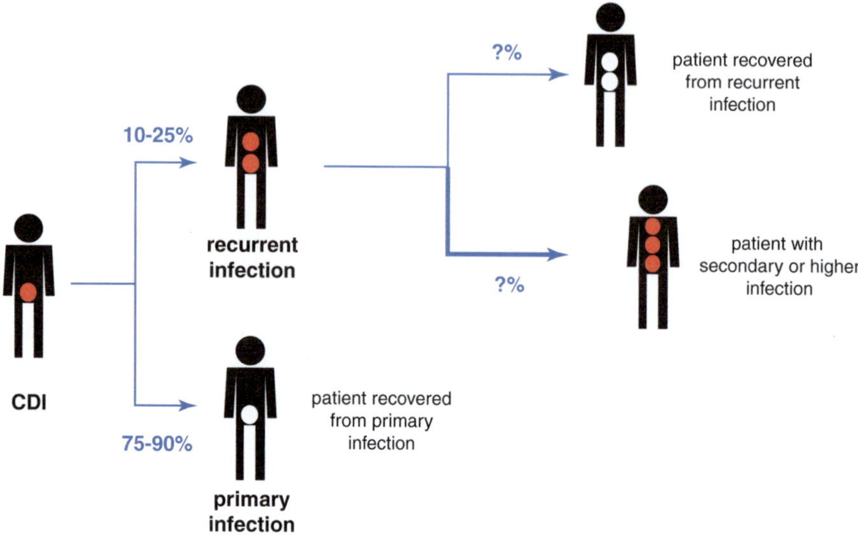



**Fig. 4.7**  Primary and recurrent CDI

neonates [18]. There is also an increase of CDI among children in community and outpatient setting [67–69], indicating that there is an epidemiologic shift with increased disease in nonhospitalized children.

## 4.7  Recurrent CDI

Recurrence (Fig. 4.7) requiring retreatment is one of the greatest challenges in the management of CDI. After effective treatment of a first CDI episode, at least one new recurrent episode occurs in 12–24% of patients [70]. The recurrence of symptoms can be the consequence of a relapse of the original infection (endogenous persistence of the same *C. difficile* strain) or reinfection (acquisition of a new strain from an exogenous source) [15].

Most studies comparing strains from the original episode and subsequent episodes (or recurrent CDAD) have found that about half are relapses and half are reinfections [71–74].

A challenge to epidemiological studies of CDI is the difficulty to discriminate relapse from reinfection on the basis of common typing schemes (e.g., ribotyping), since the same strain type can cause either reinfection or relapse. This is particular problematic in patients who are hospitalized in an environment enriched with one or more predominant strain types of *C. difficile*. In patients who had more than two prior infection episodes, the risk of further recurrence increases to 50–65% [70, 75]. Most recurrences occur within the first 30 days after completing a course of CDI therapy [76].

The cause of recurrent CDI is not completely understood, and several mechanisms have been hypothesized, including (i) new or continued disturbances in gut microbiota with subsequent loss of colonization resistance, (ii) persistence of *C. difficile* spores in the gastrointestinal tract, (iii) a defective host immune response to *C. difficile* and/or its toxins, or (iv) reinfection with a new strain [66, 77–79].

## References

1. Finney JMT. Gastro-enterostomy for cicatrizing ulcer of the pylorus. Bull Johns Hopkins Hosp. 1893;4:53.
2. Hall I, O'Toole E. Intestinal flora in newborn infants with a description of a new pathogenic anaerobe, Bacillus difficilis. Am J Dis Child. 1935;49:390.
3. Rosseau C, Poilane I, De Pontual LC, Maherault AC, Le Monnier A, Collignon A. Clostridium difficile carriage in healthy infants in the community: a potential reservoir for pathogenic strains. Clin Infect Dis. 2012;55:1209–15. https://doi.org/10.1093/cid/cis637.
4. Loo VG, Poirier L, Miller M, Oughton M, Libman MD, Michaud S, Bourgault AM, Nguyen T, Frenette C, Kelly M, Vibien A, Brassard P, Fenn S, Dewar K, Hudson TJ, Horn R, René P, Monczak Y, Dascal A. A predominately clonal multi-institutional outbreak of Clostridium difficile-associated diarrhea with high morbidity and mortality. N Engl J Med. 2005;353:2442–9. https://doi.org/10.1056/NEJMoa051639.
5. Altemeier WA, Hummel RP, Hill EO. Staphylococcal enterocolitis following antibiotic therapy. Ann Surg. 1963;157:847–57. https://doi.org/10.1097/00000658-196306000-00003.
6. Tedesco FJ, Barton RW, Alpers DH. Clindamycin-associated colitis. A prospective study. Ann Intern Med. 1974;81:429–33. https://doi.org/10.7326/0003-4819-81-4-429.
7. Bartlett JG, Onderonk AB, Cisneros RL. Clindamycin-associated colitis in hamsters: protection with vancomycin. Gasterenterology. 1977;73:772–6.
8. Larson HE, Price AB. Pseudomembranous colitis: presence of clostridial toxin. Lancet. 1977;2:1312–4. https://doi.org/10.1016/s0140-6736(77)90363-4.
9. Rifkin GD, Fekety FR, Silva J Jr. Antibiotic-induced colitis implication of a toxin neutralized by Clostridium sordellii antitoxin. Lancet. 1977;2:1103–6. https://doi.org/10.1016/s0140-6736(77)90547-5.
10. Bartlett JG, Chang TW, Gurwith M, Gorbach SL, Onderdonk AB. Antibiotic-associated pseudomembranous colitis due to toxin-producing clostridia. N Engl J Med. 1978;298:531–4. https://doi.org/10.1056/NEJM197803092981003.
11. George WL, Sutter VL, Goldstein EJ, Ludwig SL, Finegold SM. Etiology of antimicrobial-agent-associated colitis. Lancet. 1978;1:802–3. https://doi.org/10.1016/s0140-6736(78)93001-5.
12. Gerding DN. Clostridium difficile 30 years on: what has, or has not, changed and why? Int J Antimicrob Agents. 2009;33(Suppl 1):S2–8. https://doi.org/10.1016/S0924-8579(09)70008-1.
13. Larson HE, Price AB, Honour P, Borriello SP. Clostridium difficile and the etiololgy of pseudomembranous colitis. Lancet. 1978;1:1063–6. https://doi.org/10.1016/s0140-6736(78)90912-1.
14. Redelings MD, Sorvillo F, Mascola L. Increase in Clostridium difficile-related mortality rates, United States, 1999–2004. Emerg Infect Dis. 2007;13:1417–9. https://doi.org/10.3201/eid1309.061116.
15. DePestel DD, Aronoff DM. Epidemiology of Clostridium difficile infection. J Pharm Pract J Pharm Pract. 2013;26(5):464–75. https://doi.org/10.1177/0897190013499521.
16. Centers for Disease Control and Prevention (CDC). Severe Clostridium difficile-associated disease in populations previously at low risk-four states, 2005. MMWR Morb Mortal Wkly Rep. 2005;54:1201–5.
17. Centers for Disease Control and Prevention. Surveillance for community-associated Clostridium difficile – Connecticut, 2006. MMWR Morb Mortal Wkly Rep. 2008;57:340–3.

18. Kim J, Smathers SA, Prasad P, Leckerman KH, Coffin S, Zaoutis T. Epidemiological features of Clostridium difficile-associated disease among inpatients at children's hospitals in the United States, 2001–2006. Pediatrics. 2008;122:1266–70. https://doi.org/10.1542/peds.2008-0469.
19. Wilcox MH, Mooney L, Bendall R, Settle CD, Fawley WN. A case-controlled study of community-associated Clostridium difficile infection. J Antimicrob Chemother. 2008;62:388–96. https://doi.org/10.1093/jac/dkn163.
20. Lessa FC, Gould CV, McDonald LC. Current status of Clostridium difficile infection epidemiology. Clin Infect Dis. 2012;55(suppl2):S65–70. https://doi.org/10.1093/cid/cis319.
21. Khanna S, Pardi DS, Aronson SL, Kammer PP, Orenstein R, St Sauver JL, Harmsen WS, Zinsmeister AR. The epidemiology of community-acquired Clostridium difficile infection: a population-based study. Am J Gastroenterol. 2012;107:89–95. https://doi.org/10.1038/ajg.2011.398.
22. Warny M, Pépin J, Fang A, Killgore G, Thompson A, Brazier J, Frost E, McDonald LC. Toxin production by an emerging strain of Clostridium difficile associated with outbreaks of severe disease in North America and Europe. Lancet. 2005;366:1079–84. https://doi.org/10.1016/S0140-6736(05)67420-X.
23. Cookson B. Hypervirulent strains of Clostridium difficile. Postgrad Med J. 2007;83(979):291–5. https://doi.org/10.1136/pgmj.2006.056143.
24. Loo VG, Libman MD, Miller MA, Bourgault AM, Frenette CH, Kelly M, Michaud S, Nguyen T, Poirier L, Vibien A, Horn R, Laflamme PJ, René P. Clostridium difficile: a formidable foe. CMAJ. 2004;171(1):47–8. https://doi.org/10.1503/cmaj.1040836.
25. Pépin JL, Valiquette ME, Clossette B. Mortality attributed to nosocomial Clostridium difficile associated disease during an epidemic caused by a hyperviluent strain in Quebec. CMAJ. 2005a;173:1037–42. https://doi.org/10.1503/cmaj.050978.
26. Pépin JL, Saheb N, Coulombe MA, Alary ME, Corriveau MP, Authier S, Leblanc M, Rivard G, Bettez M, Primeau V, Nguyen M, Jacob CE, Lanthier L. Emergence of fluoroquinolones as the predominant risk factor for Clostridium difficile-associated diarrhea: a cohort study during an epidemic in Quebec. Clin Infect Dis. 2005b;41:1254–60. https://doi.org/10.1086/496986.
27. Poutanen SM, Simor AE. Clostridium difficile-associated diarrhea in adults. C Med Assoc J. 2004;171:51–8. https://doi.org/10.1503/cmaj.1031189.
28. McDonald LC. Clostridium difficile: responding to a new threat from an old enemy. Infect Control Hosp Epidemiol. 2005;26:672–5. https://doi.org/10.1086/502600.
29. Centers for Disease Control and Prevention (CDC). Clostridioides difficile (C. diff). https://www.cdc.gov/cdiff/index.html. Assessed 04. 06.2021.
30. McDonald LC, Killgore GE, Thompson A, Owens RC, Kazakova SV, Sambol SP, Johnson S, Gerding DN. An epidemic, toxin gene-variant strain of Clostridium difficile. N Engl J Med. 2005;353:2433–41. https://doi.org/10.1056/NEJMoa051590.
31. Goorhuis A, Van deer Kooi T, Vaessen N, Dekker FW, Van den Berg R, Harmanus C, van den Hof S, Notermans DW, Kuijper EJ. Spread and epidemiology of Clostridium difficile polymerase chain reaction ribotype 027/toxinotype III in The Netherlands. Clin Infect Dis. 2007;45:695–703. https://doi.org/10.1086/520984.
32. Muto CA, Pokrywka M, Shutt K, Mendelsohn AB, Nouri K, Posey K, Roberts T, Croyle K, Krystofiak S, Patel-Brown S, Pasculle AW, Paterson DL, Saul M, Harrison LH. A large outbreak of Clostridium difficile-associated disease with an unexpected proportion of deaths and colectomies at a teaching hospital following increased fluoroquinolone use. Infect Control Hosp Epidemiol. 2005;26:273. https://doi.org/10.1086/502539.
33. Pépin JL, Valiquette ME, Alary ME, Villemure P, Pelletier FA, Karine Pépin K, Chouinard D. Clostridium difficile-associated diarrhea in a region of Quebec from 1991–2003: a changing pattern disease severity. CMAJ. 2004;17:466–72. https://doi.org/10.1503/cmaj.1041104.
34. Bauer MP, Notermans DW, van Benthem BH, Brazier JS, Wilcox MH, Rupnik M, Monnet DL, van Dissel JT, Kuijper EJ, ECDIS Study Group. Clostridium difficile infection in Europe: a hospital-based survey. Lancet. 2011;377:63–73. https://doi.org/10.1016/S0140-6736(10)61266-4.

35. Goorhuis A, Bakker D, Corver J, Debast SB, Harmanus C, Notermans DW, Bergwerff AA, Dekker FW, Kuijper EJ. Emergence of Clostridium difficile infection due to a new hypervirulent strain, polymerase chain reaction ribotype 078. Clin Infect Dis. 2008;47:1162–70. https://doi.org/10.1086/592257.
36. Davies KA, Ashwin H, Longshaw CM, Burns DA, Davis GL, Wilcox MH, on behalf of the EUCLID study group. Diversity of Clostridium difficile PCR ribotypes in Europe: results from the European, multicentre, prospective, biannual, point-prevalence study of Clostridium difficile infection in hospitalised patients with diarrhoea (EUCLID), 2012 and 2013. Euro Surveill. **2016**;21(29):30294.
37. Eyre DW, Davies KA, Davis G, Fawley WN, Dingle KE, De Maio N, Karas A, Crook DW, Peto TEA, Walker AS, Wilcox MH, EUCLID Study Group. Two distinct patterns of Clostridium Difficile diversity across Europe indicates contrasting routes of spread. Clin Infect Dis. 2018;67:1035–44.
38. Knetsch CW, Connor TR, Mutreja A, van Dorp SM, Sanders IM, Browne HP, Harris D, Lipman L, Keessen EC, Corver J, Kuijper EJ, Lawley TD. Whole genome sequencing reveals potential spread of Clostridium difficile between humans and farm animals in the Netherlands, 2002 to 2011. Euro Surveill. 2014;19:20954. https://doi.org/10.2807/1560-7917.es2014.19.45.20954.
39. Knight DR, Squire MM, Collins DA, Riley TV. Genome analysis of Clostridium difficile PCR ribotype 014 lineage in Australian pigs and humans reveals a diverse genetic repertoire and signatures of long-range interspecies transmission. Front Microbiol. 2016;7:2138. https://doi.org/10.3389/fmicb.2016.02138.
40. Fawley WN, Freeman J, Smith C, Harmanus C, van den Berg RJ, Kuijper EJ, Wilcox MH. Use of highly discriminatory fingerprinting to analyze clusters of Clostridium difficile infection cases due to epidemic ribotype 027 strains. J Clin Microbiol. 2008;46:954–60. https://doi.org/10.1128/JCM.01764-07.
41. Borriello SP, Honour P. Concomitance of cytotoxigenic and noncytotoxigenic clostridium difficile in stool specimens. J Clin Microbiol. 1983;18:1006–7.
42. Sim JHC, Truong C, Minot SS, Greenfield N, Budvytiene I, Lohith A, Anikst V, Pourmand N, Banaei N. Determining the cause of recurrent Clostridium difficile infection using whole genome sequencing. Diagn Microbiol Infect. 2017;87:11–6. https://doi.org/10.1016/j.diagmicrobio.2016.09.023.
43. Tenover FC, Åkerlund T, Gerding DN, Goering RV, Boström T, Jonsson AM, Wong E, Wortman AT, Persing DH. Comparison of strain typing results for clostridium difficile isolates from North America. J Clin Microbiol. 2011;49:1831–7. https://doi.org/10.1128/JCM.02446-10.
44. Merrigan MM, Sambol SP, Johnson S, Gerding DN. New approach to the management of Clostridium difficile infection: colonization with non-toxigenic *C. difficile* during daily ampicillin or ceftriaxone administration. Int J Antimicrob Agents. 2009;33(Suppl. 1):S46–50. https://doi.org/10.1016/S0924-8579(09)70017-2.
45. Gerding DN, Johnson S, Rupnik M, Aktories K. Clostridium difficile binary toxin CDT: mechanism, epidemiology, and potential clinical importance. Gut Microbes. 2014;5:15–27. https://doi.org/10.4161/gmic.26854.
46. Goldenberg JZ, Yap C, Lytvyn L, Lo CKF, Beardsley J, Mertz D, Johnston BC. Probiotics for the prevention of Clostridium difficile-associated diarrhea in adults and children. Cochrane Database Syst Rev. 2017;(12):CD006095. https://doi.org/10.1002/14651858.CD006095.pub4.
47. Gerding DN, Meyer T, Lee C, Cohen SH, Murthy UK, Poirier A, Van Schooneveld TC, Pardi DS, Ramos A, Barron MA, Chen H, Villano S. Administration of spores of nontoxigenic Clostridium difficile strain M3 for prevention of recurrent C. difficile infection: a randomized clinical trial. JAMA. 2015;313:1719–27. https://doi.org/10.1001/jama.2015.3725.
48. Antonara S, Leber AL. Diagnosis of Clostridium difficile infections in children. J Clin Microbiol. 2016;54(6):1425–33. https://doi.org/10.1128/JCM.03014-15.
49. Noor A, Krilov LR. Clostridium difficile infection in children. Pediatric Annals. 2018;47(9):e359–65. https://doi.org/10.3928/19382359-20180803-01.
50. Shim JO. Clostridium difficile in children: to treat or not to treat? Pediatr Gastroenterol Hepatol Nutr. 2014;17(2):80–4. https://doi.org/10.5223/pghn.2014.17.2.80.

51. Delmée M, Verellen G, Avesani V, Francois G. Clostridium difficile in neonates: serogrouping and epidemiology. Eur J Pediatr. 1988;147:36–40. https://doi.org/10.1007/BF00442608.

52. Ellis ME, Mandal BK, Dunbar EM, Bundell KR. Clostridium difficile and its cytotoxin in infants admitted to hospital with infectious gastroenteritis. Br Med J (Clin Res Ed). 1984;288:524–6. https://doi.org/10.1136/bmj.288.6416.524.

53. Larson HE, Barclay FE, Honour P, Hill ID. Epidemiology of Clostridium difficile in infants. J Infect Dis. 1982;146(6):727–33. https://doi.org/10.1093/infdis/146.6.727.

54. Rousseau C, Lemée L, Le Monnier A, Poilane I, Pons JL, Collignon A. Prevalence and diversity of Clostridium difficile strains in infants. J Med Microbiol. 2011;60:1112–8. https://doi.org/10.1099/jmm.0.029736-0.

55. Toma S, Lesiak G, Magus M, Lo HL, Delmée M. Serotyping of Clostridium difficile. J Clin Microbiol. 1988;26:426–8.

56. Boenning DA, Fleisher GR, Campos JM, Hulkower CW, Quinlan RW. Clostridium difficile in a pediatric outpatient population. Pediatr Infect Dis. 1982;1(5):336–8. https://doi.org/10.1097/00006454-198209000-00011.

57. Jangi S, Lamont JT. Asymptomatic colonization by Clostridium difficile in infants: implications for disease in later life. J Pediatr Gastroenterol Nutr. 2010;51(1):2–7. https://doi.org/10.1097/MPG.0b013e3181d29767.

58. Richardson SA, Alcock PA, Gray J. Clostridium difficile and its toxin in healthy neonates. Br Med J (Clin Res Ed). 1983;287(6396):878. https://doi.org/10.1136/bmj.287.6396.878.

59. Viscidi R, Willey S, Bartlett JG. Isolation rates and toxigenic potential of Clostridium difficile isolates from various patient populations. Gastroenterology. 1981;81(1):5–9.

60. Benno Y, Sawada K, Mitsuoka T. The intestinal microflora of infants: composition of fecal flora in breast-fed and bottle-fed infants. Microbiol Immunol. 1984;28(9):975–86. https://doi.org/10.1111/j.1348-0421.1984.tb00754.x.

61. Pothoulakis C, Lamont JT. Microbes and microbial toxins: paradigms for microbialmucosal interactions II. The integrated response of the intestine to Clostridium difficile toxins. Am J Physiol Gastrointest Liver Physiol. 2001;280(2):G178–83. https://doi.org/10.1152/ajpgi.2001.280.2.G178.

62. Vesikari T, Isolauri E, Mäki M, Grönroos P. Clostridium difficile in young children. Association with antibiotic usage. Acta Paediatr Scand. 1984;73(1):86–91. https://doi.org/10.1111/j.1651-2227.1984.tb09903.x.

63. Nylund CM, Goudie A, Garza JM, Fairbrother G, Cohen MB. Clostridium difficile infection in hospitalized children in the United States. Arch Pediatr Adolesc Med. 2011;165(5):451–7.

64. Schutze GE, Willoughby RE. Committee on Infectious Diseases; American Academy of Pediatrics. Clostridium difficile infection in infants and children. Pediatrics. 2013;131(1):196–200. https://doi.org/10.1542/peds.2012-2992.

65. Zilberberg MD, Tillotson GS, McDonald C. Clostridium difficile infections among hospitalized children, United States, 1997–2006. Emerg Infect Dis. 2010;16(4):604–9.

66. McDonald LC, Gerding DN, Johnson S, Bakken JS, Carroll KC, Coffin SE, Dubberke ER, Garey KW, Gould CV, Kelly C, Loo V, Sammons JS, Sandora TJ, Wilcox MH. Clinical Practice Guidelines for Clostridium Difficile Infection in Adults and Children: 2017 Update by the Infectious Diseases Society of America (IDSA) and Society for Healthcare Epidemiology of America (SHEA). Clin Infect Dis. 2018;66(7):e1–e48. https://doi.org/10.1093/cid/cix1085.

67. Baker SS, Faden H, Sayej W, Patel R, Baker RD. Increasing incidence of community associated atypical Clostridium difficile disease in children. Clin Pediatr (Phila). 2010;49:644–7. https://doi.org/10.1177/0009922809360927.

68. Benson L, Song X, Campos J, Singh N. Changing epidemiology of Clostridium difficile-associated disease in children. Infect Control Hosp Epidemiol. 2007;28:1233–5. https://doi.org/10.1086/520732.

69. Khanna S, Baddour LM, Huskins WC, Kammer PP, Faubion WA, Zinsmeister AR, Harmsen WS, Pardi DS. The epidemiology of Clostridium difficile infection in children: a population-based study. Clin Infect Dis. 2013;56:1401–6. https://doi.org/10.1093/cid/cit075.

70. McFarland LV, Elmer GW, Surawicz CM. Breaking the cycle: treatment strategies for 163 cases of recurrent Clostridium difficile disease. Am J Gastroenterol. 2002;97:1769–75. https://doi.org/10.1111/j.1572-0241.2002.05839.x.

71. Barbut F, Richard A, Hamadi K, Chomette V, Burghoffer B, Petit JC. Epidemiology of recurrences or reinfections of Clostridium difficile-associated diarrhea. J Clin Microbiol. 2000;38:2386–8.

72. Johnson S, Adelmann A, Clabots CR, Peterson LR, Gerding DN. Recurrences of Clostridium difficile diarrhea not caused by the original infecting organism. J Infect Dis. 1989;159:340–3. https://doi.org/10.1093/infdis/159.2.340.

73. Tang-Feldman Y, Mayo S, Silva J, Cohen SH. Molecular analysis of Clostridium difficile strains isolated from 18 cases of recurrent Clostridium difficile-associated diarrhea. J Clin Microbiol. 2003;41:3413–4. https://doi.org/10.1128/JCM.41.7.3413-3414.2003.

74. Wilcox MH, Fawley WN, Settle CD, Davidson A. Recurrence of symptoms in Clostridium difficile infection – relapse or reinfection? J Hosp Infect. 1998;38:93–100. https://doi.org/10.1016/S0195-6701(98)90062-7.

75. Vincent Y, Manji A, Gregory-Miller K, Lee C. A review of management of Clostridium difficile infection: primary and recurrence. Antibiotics (Basel). 2015;4(4):411–23. https://doi.org/10.3390/antibiotics4040411.

76. Kelly CP. Can we identify patients at high risk of recurrent Clostridium difficile infection? Clin Microbiol Infec. 2012;18(Suppl 6):21–7. https://doi.org/10.1111/1469-0691.12046.

77. Drekonja DM, Amundson WH, Decarolis DD, Kuskowski MA, Lederle FA, Johnson JR. Antimicrobial use and risk for recurrent Clostridium difficile infection. Am J Med. 2011;124(11):1081.e1–1081.e10817. https://doi.org/10.1016/j.amjmed.2011.05.032.

78. Garey KW, Sethi S, Yadav Y, DuPont HL. Meta-analysis to assess risk factors for recurrent Clostridium difficile infection. J Hosp Infect. 2008;70(4):298–304.

79. Johnson S. Recurrent Clostridium difficile infection: a review of risk factors, treatments, and outcomes. J Infect. 2009;58(6):403–10. https://doi.org/10.1016/j.jinf.2009.03.010.

# Risk Factors for CDI

5

## Contents



Several important patient risk factors for CDI have been identified and are summarized in Fig. 5.1.

## 5.1 Exposure to Healthcare Institutions

A significant patient risk factor for CDI is recent healthcare exposure [1–4]. Frequent hospitalization and increased length of staying in a hospital are associated with an increased risk for CDI [1]. Patients are exposed to *C. difficile* spores through contact with the healthcare environment or healthcare workers hands, stressing the importance of infection control and prevention efforts in all healthcare settings.


© The Author(s), under exclusive license to Springer Nature Switzerland AG 2021

H. Sommermeyer, J. Piątek, *Clostridioides difficile*,
https://doi.org/10.1007/978-3-030-81100-6_5

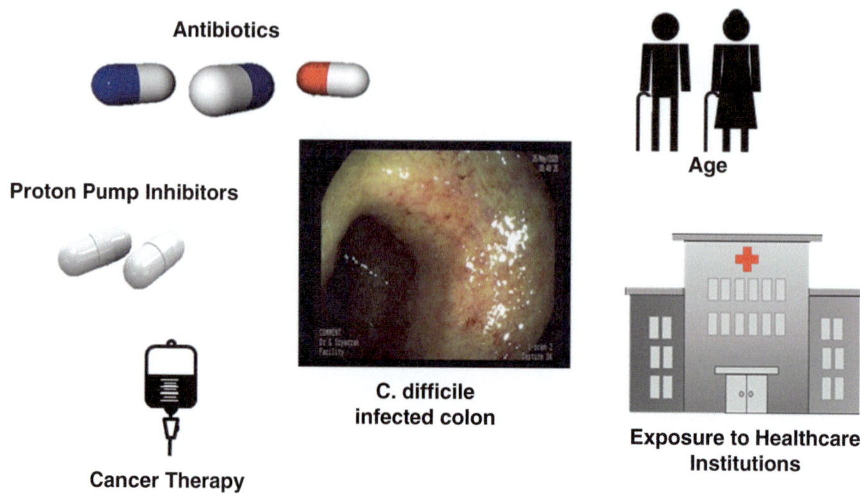



**Fig. 5.1** Overview of CDI risk factors

## 5.2 Antibiotics

Antibacterial exposure is a dominant risk for the development of CDI. Nearly every antibacterial has been associated with the development of CDI, including metronidazole and vancomycin [5]. The indigenous gut microbiota functions to protect against colonization or infection by pathogenic organisms, including *C. difficile*; this is known as "colonization resistance" [6]. Every antibiotic therapy can disrupt the competitive balance or homeostasis in the gut microbiota and promote the overgrowth of *C. difficile*, serving as major step in the development of CDI [6]. Recurrent CDI after treatment with vancomycin or metronidazole, to some extent, is thought to be related to the ability of these antibiotics to disrupt the gut microbiota [7–9]. These therapies inhibit not only *C. difficile* but also other bacterial communities, thereby leading to loss of colonization resistance [7–9]. In some instances, this allows germination of *C. difficile* spores and proliferation of vegetative cells in the gastrointestinal tract, causing a re-emergence of *C. difficile* populations leading to recurrence [9].

Several antibacterial classes appear to increase the risk compared to others (Table 5.1), including clindamycin, cephalosporins, and fluoroquinolones [10, 11].

The relative risk of a particular antibiotic and its association with CDI depends on the local prevalence of strains that are highly resistant to a given antimicrobial agent [3]. For instance, clindamycin is associated with a high risk for CDI, particularly with the "J strain" (REA type J7/9) which is resistant to clindamycin and was responsible for large outbreaks in the early 1990s [12]. More recently, fluoroquinolone use has been related to outbreaks caused by the ribotype 027 strain, corresponding with a marked increase in fluoroquinolone resistance in this strain [13,

**Table 5.1**  Antibiotics related to CDI

| Very commonly related | Less commonly related | Uncommonly related |
|---|---|---|
| Clindamycin | Sulfonamides | Aminoglycosides |
| Ampicillin | Macrolides | Rifampin |
| Amoxicyllin | Carbapenems | Tetracycline |
| Cephalosporines | Other penicillins | Chloramphicol |
| Quinolones (moxifloxacin & gatifloxacin) | | |

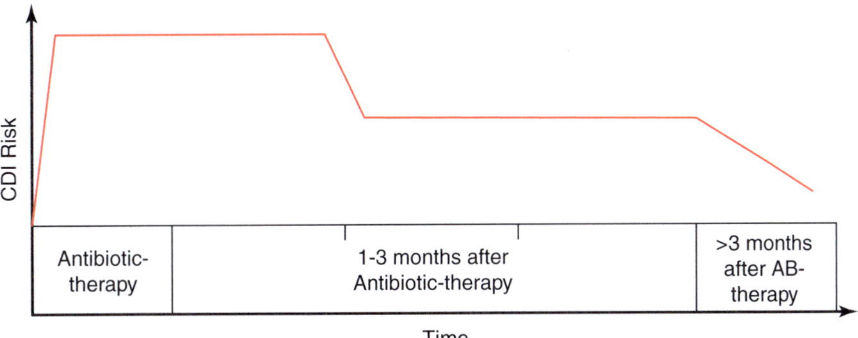


**Fig. 5.2**  Time dependence of CDI risk during and after antibiotic therapy

14]. Cumulative antibiotic exposure via dose, use of multiple antibacterial agents, and increased days of antibiotic exposure all contribute to the risk of CDI [11, 15]. Alternatively, limited exposure such as a single-dose antibacterial exposure for surgical prophylaxis also increases the risk of both *C. difficile* colonization and infection [3, 16, 17]. It has been hypothesized that the increased incidence of CDI in young, healthy peripartum women may be a result of exposure to antibacterial prophylaxis associated with cesarean sections [18].

Exposure to antibiotics increases the risk for CDI not only during receipt of antibiotic therapy but also in the 3 months after cessation of therapy, with the highest risk during the first month after antibacterial therapy (Fig. 5.2) [10].

Importantly, administration of antibacterials during and after treatment of CDI has been associated with lower cure rates, prolonged time to diarrhea resolution, and a trend toward recurrent CDI [8]. A recent retrospective study demonstrated administration of antibiotics within 30 days after CDI therapy was significantly associated with the 90-day risk of recurrence [17]. Therefore, antimicrobial stewardship and the judicious use of antibiotics are of great importance to control CDI. Unfortunately, in a recent study it was found that up to 26% of patients with a current or a recent history of CDI receive unnecessary and potentially avoidable antibacterial therapy [19]. Antimicrobial stewardship efforts should be an essential component of any intervention aimed at reducing the risk of CDI.

## 5.3    Proton Pump Inhibitors

The administration of gastric acid suppression agents, such as proton pump inhibitors (PPIs) and histamine-2 receptor antagonists (H2RAs), has been associated with an increased risk of CDI [3, 20–22]. While the precise biological basis for this association is not fully understood, several hypotheses have been investigated. Decreased gastric acidity could lead to an inadequate sterilization of ingested pathogen organisms, including *C. difficile*, resulting in increased colonization of the upper gastrointestinal tract [23, 24]. Acid-suppressing agents may also contribute to disruption of the indigenous gut microbiota, resulting in bacterial colonization of the stomach or upper small intestine, thereby promoting an environment conducive for *C. difficile* [25]. Bile salts play an important role in stimulating the conversion of *C. difficile* spores to vegetative cells, and alterations of the gut microbiota composition have been demonstrated to markedly increase the luminal concentrations of unconjugated bile acids [23, 26]. Lastly, studies also suggest that PPIs may have an effect on intestinal neutrophil activity, which may interfere with defense mechanisms protective against CDI [27–29].

The U.S. Food and Drug Administration issued a drug safety communication noting a possible association between the use of PPIs and CDI [22]. The FDA communication was based on the analysis of 28 observational studies, 23 of which demonstrated a higher risk of CDI associated with PPI exposure. Most of the studies reviewed found the risk of CDI ranged from 1.4 to 2.75-fold higher among those exposed to PPIs [22]. Two meta-analyses both found an overall increased risk of CDI with PPI use [20, 21]. Concomitant use of PPIs and antibiotics conferred an even greater risk than either treatment alone [21]. H2RAs appeared to carry a lower risk compared with PPIs [21].

It is important to note that the effectiveness of PPIs for the treatment of upper gastrointestinal disorders has led to their overutilization in multiple treatment arenas, exposing patients to an increasing number of potential risks [30]. Due to the potentially increased risk of CDI, the rational and prudent use of acid-suppressing agents should be encouraged.

## 5.4    Advanced Age as Risk Factor for CDI

Advanced age is an important CDI risk factor as evident by the disproportionate high incidence of CDI in older patients. The particular reasons for this association are likely multifactorial, including diminished immune response, a diminished variety of bacteria in the gut, increased comorbidities and increased exposure to antibiotics and the healthcare environment [2, 3].

## 5.5   Cancer Therapy as Risk Factor for CDI

Cancer chemotherapy has also been shown to be a CDI risk factor, independent of antibiotic exposure [3, 31]. Several chemotherapeutic agents have been implicated, including carboplatin, cisplatin, cyclophosphamide, doxorubicin, methotrexate, topotecan, paclitaxel, vinorelbine, and 5-fluorouracil, as well as others [31]. The mechanism is thought to be a result of the antimicrobial activity of the chemotherapeutic agent and subsequent alterations in gut microbiota [3, 31]. Additional possible mechanisms include severe inflammatory changes induced by chemotherapy, intestinal necrosis promoting an anaerobic environment within the gastrointestinal tract, decreased degradation of *C. difficile* toxins, and delayed reestablishment of indigenous microbiota [31].

## 5.6   Other Risk Factors for CDI

The patient's ability to produce an adequate immune response may be a key risk for CDI development and recurrence [2, 32–34]. In adults, the ability to produce antitoxin antibodies may be an important determinant leading to colonization versus active infection [33–35]. Individuals that fail to produce sufficient quantities of immunoglobulin G (IgG) directed against toxin A are more likely to develop CDI and may also be at higher risk for recurrences [33, 34]. Other investigations have demonstrated that patients with a common single-nucleoside polymorphism in the −251 region of the interleukin (IL)-8 gene promoter are at increased risk for CDI [36].

Populations with underlying chronic comorbid conditions, including those with chronic kidney disease [37], human immunodeficiency virus (HIV) [3, 38], solid organ [39, 40] or hematopoietic stem cell transplantation [31], and inflammatory bowel disease [41] all appear to be at an increased risk of developing CDI. It is not entirely clear if these patients are at specific increased risk because of their underlying immunosuppression, exposure to specific medications, frequent exposure to antimicrobials, frequent and prolonged exposure to the healthcare environment, or some combination of these factors.

## 5.7   Risk Factors for Community-Acquired CDI

Risk factors for CA-CDI have not been fully elucidated [42]. While increased exposure to antibacterial agents could be an important factor, there have been reports of CA-CDI in the absence of antibacterial exposure [43, 44]. As discussed, the use of gastric acid-suppressive agents could also be a potential risk factor. Interestingly, a recent study demonstrated the odds of developing CDI were 80% greater in current smokers and 33% greater in former smokers compared to nonsmokers [45]. The study included those >50 years of age and CDI diagnosed both in the inpatient and outpatient setting. Increased transmission of *C. difficile* in the community could play a role. *C. difficile* has been isolated from the environment in outpatient healthcare areas and daycare centers, possibly serving as a reservoir for transmission [42]. It has been suggested that exposure to households with children less than 2 years of age, a group that is known to be highly colonized with *C. difficile*, could serve as a potential reservoir for *C. difficile* transmission [46–48]. Lastly, it has been hypothesized that retail food products and domestic animals could serve as potential sources of *C. difficile* exposure [49].

## 5.8   Risk Factors for Recurrent CDI

Recurrent CDI represents one of the greatest challenges in the management of CDI. Risk factors for recurrence (Fig. 5.3) and those for an initial episode are not necessarily mutually exclusive. Several risk factors that initially predispose a patient to CDI remain an ongoing problem. A predominant risk factor is an episode of initial recurrence [50]. Studies have demonstrated that the risk of recurrence more than doubles after two or more recurrences [51, 52]. Individuals that lack a host response to produce sufficient quantities of antibodies directed against toxin A and those individuals with a common single-nucleoside polymorphism in the −251 region of the interleukin-8 gene promoter may have an increased risk for recurrences [32–34]. The risk of recurrences increases with advanced age with those patients ≥65 year of age at highest risk [2, 7, 53]. Disruption of the gut microbiota and loss of

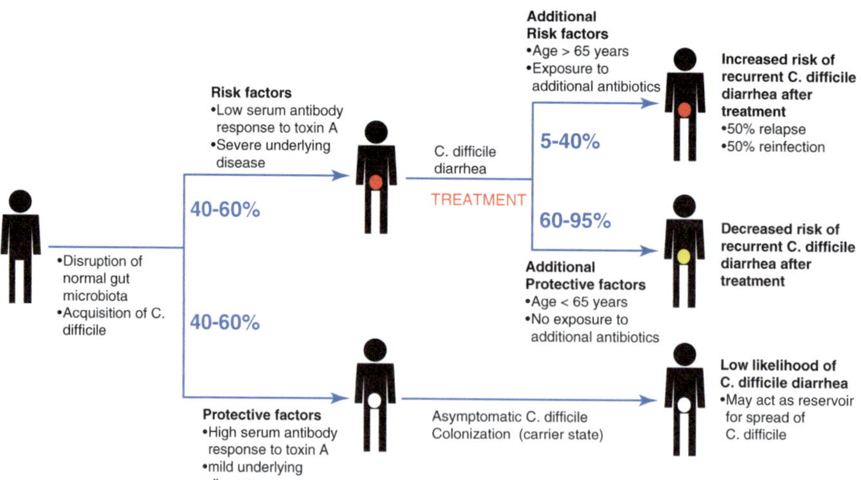


**Fig. 5.3**   Risk factors of recurrent CDI

colonization resistance has been investigated as a risk factor of recurrence [6]. Other risk factors for recurrence include concurrent use of antibiotics for non-CDI, acid-suppressing agents, exposure to the healthcare environment, and underlying chronic comorbidities [2, 7, 8, 17, 53].

## 5.9  CDI Risk Factors in Children

Many of the risk factors for *C. difficile* infection in children mirror those for adults, including recent antibiotic exposure, hospitalization, and underlying complex chronic conditions such as malignancy, solid organ transplant, and inflammatory bowel disease [54–59]. In children, the presence of a gastrostomy or jejunostomy tube has been found to be an additional independent risk factor [57]. Recent studies suggest that acid-suppressing medications may also be an independent risk factor for CDI in children, although the association has been more consistently observed in children who receive histamine-2 receptor antagonists than PPIs [60, 61].

## References


1. Freeman J, Bauer MP, Baines SD, Corver J, Fawley WN, Goorhuis B, Kuijper EJ, Wilcox MH. The changing epidemiology of Clostridium difficile infections. Clin Microbiol Rev. 2010;23:529–49. https://doi.org/10.1128/CMR.00082-09.
2. Johnson S. Recurrent Clostridium difficile infection: a review of risk factors, treatments, and outcomes. J Infect. 2009;58(6):403–10. https://doi.org/10.1016/j.jinf.2009.03.010.
3. McDonald LC, Gerding DN, Johnson S, Bakken JS, Carroll KC, Coffin SE, Dubberke ER, Garey KW, Gould CV, Kelly C, Loo V, Sammons JS, Sandora TJ, Wilcox MH. Clinical Practice Guidelines for Clostridium Difficile Infection in Adults and Children: 2017 Update by the Infectious Diseases Society of America (IDSA) and Society for Healthcare Epidemiology of America (SHEA). Clin Infect Dis. 2018;66(7):e1–e48. https://doi.org/10.1093/cid/cix1085.
4. Pacheco SM, Johnson S. Important clinical advances in the understanding of Clostridium difficile infection. Curr Opin Gastroenterol. 2013;29:42–8. https://doi.org/10.1097/MOG.0b013e32835a68d4.
5. Kelly CP, Pothoulakis C, LaMont JT. Clostridium difficile colitis. N Engl J Med. 1994;330:257. https://doi.org/10.1056/NEJM199401273300406.
6. Chang JY, Antonopoulos DA, Kalra A, Adriano Tonelli A, Khalife WT, Schmidt TM, Young VB. Decreased diversity of the fecal microbiome in recurrent Clostridium difficile-associated diarrhea. J Infect Dis. 2008;197:435–8. https://doi.org/10.1086/525047.
7. Kelly CP. Can we identify patients at high risk of recurrent Clostridium difficile infection? Clin Microbiol Infect. 2012;18(6):21–7. https://doi.org/10.1111/1469-0691.12046.
8. Mullane KM, Miller MA, Weiss K, Lentnek A, Golan Y, Sears PS, Shue YK, Louie TJ, Gorbach SL. Efficacy of fidaxomicin versus vancomycin as therapy for Clostridium difficile infection in individuals taking concomitant antibiotics for other concurrent infections. Clin Infect Dis. 2011;53:440–7. https://doi.org/10.1093/cid/cir404.
9. Tannock GW, Munro K, Taylor C, Lawley B, Young W, Byrne B, Emery J, Louie T. A new macrocyclic antibiotic, fidaxomicin (OPT-80), causes less alterations to the bowel microbiota of Clostridium difficile-infected patients than does vancomycin. Microbiology. 2010;156:3354–9. https://doi.org/10.1099/mic.0.042010-0.

10. Hensgens MP, Goorhuis A, Dekkers OM, Kuijper EJ. Time interval of increased risk for Clostridium difficile infection after exposure to antibiotics. J Antimicrob Chemother. 2012;67(3):742–8. https://doi.org/10.1093/jac/dkr508.

11. Stevens V, Dumyati G, Fine LS, Fisher SG, van Wijngaarden E. Cumulative antibiotic exposures over time and the risk of Clostridium difficile infection. Clin Infect Dis. 2011;53:42–8. https://doi.org/10.1093/cid/cir301.

12. Johnson S, Samore MH, Farrow KA, Killgore GE, Tenover FC, Lyras D, Rood JI, DeGirolami P, Baltch AL, Rafferty ME, Pear SM, Gerding DN. Epidemics of diarrhea caused by a clindamycin-resistant strain of Clostridium difficile in four hospitals. N Engl J Med. 1999;341:1645. https://doi.org/10.1056/NEJM199911253412203.

13. McDonald LC. Clostridium difficile: responding to a new threat from an old enemy. Infect Control Hosp Epidemiol. 2005;26:672–5. https://doi.org/10.1086/502600.

14. Muto CA, Pokrywka M, Shutt K, Mendelsohn AB, Nouri K, Posey K, Roberts T, Croyle K, Krystofiak S, Patel-Brown S, Pasculle AW, Paterson DL, Saul M, Harrison LH. A large outbreak of Clostridium difficile-associated disease with an unexpected proportion of deaths and colectomies at a teaching hospital following increased fluoroquinolone use. Infect Control Hosp Epidemiol. 2005;26:273. https://doi.org/10.1086/502539.

15. Bartlett JG, Gerding DN. Clinical recognition and diagnosis of Clostridium difficile infection. Clin Infect Dis. 2008;46(Suppl 1):S12–8. https://doi.org/10.1086/521863.

16. Carignan A, Allard C, Pépin J, Cossette B, Nault V, Valiquette L. Risk of Clostridium difficile infection after perioperative antibacterial prophylaxis before and during an outbreak of infection due to a hypervirulent strain. Clin Infect Dis. 2008;46:1838–43. https://doi.org/10.1086/588291.

17. Drekonja DM, Amundson WH, Decarolis DD, Kuskowski MA, Lederle FA, Johnson JR. Antimicrobial use and risk for recurrent Clostridium difficile infection. Am J Med. 2011;124(11):1081.e1–1081.e10817. https://doi.org/10.1016/j.amjmed.2011.05.032.

18. Kuntz JL, Yang M, Cavanaugh J, Saftlas A, Polgreen P. Trend in Clostridium difficile infection among peripartum women. Infect Control Hosp Epidemiol. 2010;31:532–4. https://doi.org/10.1086/652454.

19. Shaughnessy MK, Amundson WH, Kuskowski MA, DeCarolis DD. Unnecessary antimicrobial use in patients with current or recent Clostridium difficile infection. Infect Control Hosp Epidemiol. 2013;34:109–16. https://doi.org/10.1086/669089.

20. Janarthanan S, Ditah I, Adler DG, Ehrinpreis M. Clostridium difficile-associated diarrhea and proton pump inhibitor therapy: a meta-analysis. Am J Gastroenterol. 2012;107:1001–10. https://doi.org/10.1038/ajg.2012.179.

21. Kwok CS, Arthur AK, Anibueze CI, Singh S, Cavallazzi R, Loke YK. Risk of Clostridium difficile infection with acid suppressing drugs and antibiotics: meta-analysis. Am J Gastroenterol. 2012;107:1011–9. http://www.nature.com/ajg/journal/v107/n7/abs/ajg2012108a.html

22. United States Department of Health and Human Services. FDA Drug Safety Communication (02-08-2012): Clostridium difficile-associated diarrhea can be associated with stomach acid drugs known as proton pump inhibitors (PPIs). 2012. http://www.fda.gov/drugs/drugsafety/ucm290510.htm. Accessed 04/06/2021.

23. Jump RL, Pultz MJ, Donskey CJ. Vegatative Clostridium difficile survives in room air on moist surfaces and in gastric contents with reduced acidity: a potential mechanism to explain the association between proton pump inhibitors and C. difficile-associated diarrhea? Antimicrob Agents Chemother. 2007;51:2883–7. https://doi.org/10.1128/AAC.01443-06.

24. Thorens J, Froehlich F, Schwizer W, Saraga E, Bille J, Gyr K, Duroux P, Nicolet M, Pignatelli B, Blum AL, Gonvers JJ, Fried M. Bacterial overgrowth during treatment with omeprazole compared with cimetidine: a prospective randomized double blind study. Gut. 1996;39:54–9. https://doi.org/10.1136/gut.39.1.54.

25. Williams C. Occurrence and significance of gastric colonization during acid-inhibitory therapy. Best Pract Res Clin Gastroenterol. 2001;15:511–21. https://doi.org/10.1053/bega.2001.0191.

26. Wilson KH, Sheagren JN, Freter R. Population dynamics of ingested Clostridium difficile in the gastrointestinal tract of the Syrian hamster. J Infect Dis. 1985;151:355–61. https://doi.org/10.1093/infdis/151.2.355.
27. Agastya G, West BC, Callahan JM. Omperazole inhibits phagocytosis and acidification of phagolysosomes of normal human neutrophils in vitro. Immunopharmacol Immunotoxicol. 2000;22:357–72. https://doi.org/10.3109/08923970009016425.
28. Yoshida N, Yoshikawa T, Tanaka Y, Fujita N, Kassai K, Naito Y, Kondo M. A new mechanism for anti-inflammatory actions of proton pump inhibitors—inhibitory effects on neutrophil-endothelial cell interactions. Aliment Pharmacol Ther. 2004;14(Suppl 1):74–8. https://doi.org/10.1046/j.1365-2036.2000.014s1074.x.
29. Zedtwitz-Liebenstein K, Wenisch C, Patruta S, Parschalk B, Daxböck F, Graninger W. Omperazole treatment diminishes intra- and extracellular neutrophil reactive oxygen production and bactericidal activity. Crit Care Med. 2002;30:1118–22. https://doi.org/10.1097/00003246-200205000-00026.
30. Heidelbaugh JJ, Kim AH, Chang R, Walker PC. Overutilization of proton-pump inhibitors: what the clinician needs to know. Ther Adv Gastroenterol. 2012;5:219–32. https://doi.org/10.1177/1756283X12437358.
31. Chopra T, Alangaden GJ, Chandrasekar P. Clostridium difficile infection in cancer patients and hematopoietic stem cell transplant recipients. Expert Rev Anti-Infect Ther. 2010;8:1113–9. https://doi.org/10.1586/eri.10.95.
32. Garey KW, Jiang Z, Ghantoji S, Tam VH, Arora V, DuPont HL. A common polymorphism in the interleukin-8 gene promoter is associated with an increased risk for recurrent Clostridium difficile infection. Clin Infect Dis. 2010;51:1406–10. https://doi.org/10.1086/657398.
33. Kyne L, Warny M, Qamar A, Kelly CP. Asymptomatic carriage of Clostridium difficile and serum levels of IgG antibody against toxin A. N Engl J Med. 2000;342:390–7. https://doi.org/10.1056/NEJM200002103420604.
34. Kyne L, Warny M, Qamar A, Kelly CP. Association between antibody response to toxin A and protection against recurrent Clostridium difficile diarrhea. Lancet. 2001;357(9251):189–93. https://doi.org/10.1016/S0140-6736(00)03592-3.
35. Rupnik M, Wilcox MH, Gerding DN. Clostridium difficile infection: new developments in epidemiology and pathogenesis. Nat Rev Microbiol. 2009;7:526–36.
36. Jiang ZD, DuPont HL, Garey K, Price M, Graham G, Okhuysen P, Dao-Tran T, LaRocco M. A common polymorphism in the interleukin 8 gene promoter is associated with Clostridium difficile diarrhea. Am J Gastroenterol. 2006;101(5):1112–6. https://doi.org/10.1111/j.1572-0241.2006.00482.x.
37. Eddi R, Malik MN, Shakrov R, Baddoura WJ, Chandran C, Debari VA. Chronic kidney disease as a risk factor for Clostridium difficile infection. Nephrology. 2010;15:471–5. https://doi.org/10.1111/j.1440-1797.2009.01274.x.
38. Raines DL, Lopez FA. Clostridium difficile infection in non-HIV-immunocompromised patients and in HIV-infected patients. Curr Gastroenterol Rep. 2011;13(4):344–50. https://doi.org/10.1007/s11894-011-0196-6.
39. Boutros M, Al-Shaibi M, Chan G, Cantarovich M, Rahme E, Paraskevas S, Deschenes M, Ghali P, Wong P, Fernandez M, Giannetti N, Cecere R, Hassanain M, Chaudhury P, Metrakos P, Tchervenkov J, Barkun JS. Clostridium difficile colitis: increasing incidence, risk factors, and outcomes in solid organ transplant recipients. Transplantation. 2012;93:1051–7. https://doi.org/10.1097/TP.0b013e31824d34de.
40. Dubberke ER, Riddle DJ. Clostridium difficile in solid organ transplant recipients. Am J Transplant. 2009;9(Suppl 4):s35–40. https://doi.org/10.1097/MOT.0000000000000430.
41. Navaneethan U, Venkatesh PGK, Shen B. Clostridium difficile infection in inflammatory bowel disease—understanding the evolving relationship. World J Gastroenterol. 2010;16:4892–904.
42. Lessa FC, Gould CV, McDonald LC. Current status of Clostridium difficile infection epidemiology. Clin Infect Dis. 2012;55(Suppl 2):S65–70. https://doi.org/10.1093/cid/cis319.

43. Dumyati G, Stevens V, Hannett GE, Thompson AD, Long C, MacCannell D, Limbago B. Community-associated Clostridium difficile infections, Monroe county, New York, USA. Emerg Infect Dis. 2012;18:392–400. https://doi.org/10.3201/eid1803.102023.

44. Kutty PK, Wood CW, Sena AC, Benoit SR, Naggie S, Frederick J, Evans S, Engel J, McDonald LC. Risk factors for and estimated incidence of community associated Clostridium difficile infection, North Carolina, USA. Emerg Infect Dis. 2010;16:197–204. https://doi.org/10.3201/eid1602.090953.

45. Rogers MA, Greene MT, Saint S, Chenoweth CE, Malani PN, Trivedi I, Aronoff DM. Higher rates of Clostridium difficile infection among smokers. PLoS One. 2012;7:e42091. https://doi.org/10.1371/journal.pone.0042091.

46. Pépin J, Gonazales M, Valiquette L. Risk of secondary cases of Clostridium difficile infection among household contacts of index cases. J Infect Dis. 2012;64:387–90. https://doi.org/10.1016/j.jinf.2011.12.011.

47. Rosseau C, Poilane I, De Pontual LC, Maherault AC, Le Monnier A, Collignon A. Clostridium difficile carriage in healthy infants in the community: a potential reservoir for pathogenic strains. Clin Infect Dis. 2012;55:1209–15. https://doi.org/10.1093/cid/cis637.

48. Wilcox MH, Mooney L, Bendall R, Settle CD, Fawley WN. A case-controlled study of community-associated Clostridium difficile infection. J Antimicrob Chemother. 2008;62:388–96. https://doi.org/10.1093/jac/dkn163.

49. Gould LH, Limbago B. Clostridium difficile in food and domestic animals: a new foodborne pathogen? Clin Infect Dis. 2010;51:577–82. https://doi.org/10.1086/655692.

50. Bauer MP, Notermans DW, van Benthem BH, Brazier JS, Wilcox MH, Rupnik M, Monnet DL, van Dissel JT, Kuijper EJ, ECDIS Study Group. Clostridium difficile infection in Europe: a hospital-based survey. Lancet. 2011;377:63–73. https://doi.org/10.1016/S0140-6736(10)61266-4.

51. McFarland LV, Surawicz CM, Greenberg RN, Fekety R, Elmer GW, Moyer KA, Melcher SA, Bowen KE, Cox JL, Noorani Z. A randomized placebo controlled trial of Saccharomyces boulardii in combination with standard antibiotics for Clostridium difficile disease. JAMA. 1994;271:1913–8. https://doi.org/10.1001/jama.1994.03510480037031.

52. McFarland LV, Elmer GW, Surawicz CM. Breaking the cycle: treatment strategies for 163 cases of recurrent Clostridium difficile disease. Am J Gastroenterol. 2002;97:1769–75. https://doi.org/10.1111/j.1572-0241.2002.05839.x.

53. Garey KW, Sethi S, Yadav Y, DuPont HL. Meta-analysis to assess risk factors for recurrent Clostridium difficile infection. J Hosp Infect. 2008;70(4):298–304.

54. Kim J, Smathers SA, Prasad P, Leckerman KH, Coffin S, Zaoutis T. Epidemiological features of Clostridium difficile-associated disease among inpatients at children's hospitals in the United States, 2001–2006. Pediatrics. 2008;122:1266–70. https://doi.org/10.1542/peds.2008-0469.

55. Nylund CM, Goudie A, Garza JM, Fairbrother G, Cohen MB. Clostridium difficile infection in hospitalized children in the United States. Arch Pediatr Adolesc Med. 2011;165(5):451–7.

56. Pant C, Anderson MP, Deshpande A, Altaf MA, Grunow JE, Atreja A, Sferra TJ. Health care burden of Clostridium difficile infection in hospitalized children with inflammatory bowel disease. Inflamm Bowel Dis. 2013;19:1080–5. https://doi.org/10.1097/MIB.0b013e3182807563.

57. Sandora TJ, Fung M, Flaherty K, Helsing L, Scanlon P, Potter-Bynoe G, Courtney A, Lee G. Epidemiology and risk factors for Clostridium difficile infection in children. Pediatr Infect Dis J. 2011;30:580–4. https://doi.org/10.1097/INF.0b013e31820bfb29.

58. Tai E, Richardson LC, Townsend J, Howard E, Mcdonald LC. Clostridium difficile infection among children with cancer. Pediatr Infect Dis J. 2011;30:610–2. https://doi.org/10.1097/INF.0b013e31820970d1.

59. Thompson CM Jr, Gilligan PH, Fisher MC, Long SS. Clostridium difficile cytotoxin in a pediatric population. Am J Dis Child. 1983;137:271–4. https://doi.org/10.1001/archpedi.1983.02140290057015.

60. Brown KE, Knoderer CA, Nichols KR, Crumby AS. Acid-suppressing agents and risk for Clostridium difficile infection in pediatric patients. Clin Pediatr (Phila). 2015;54:1102–6. https://doi.org/10.1177/0009922815569201.
61. Nylund CM, Eide M, Gorman GH. Association of Clostridium difficile infections with acid suppression medications in children. J Pediatr. 2014;165:979–84.e1. https://doi.org/10.1016/j.jpeds.2014.06.062.

# Diagnosis of CDI

**6**

## Contents



The diagnosis of CDI [1] should be based on clinical signs (diarrhea) and on detection of the toxin-producing *C. difficile* in the stool sample by different microbiological procedures. Loose stool samples taking the shape of the container indicate that the patient is a symptomatic carrier of toxigenic *C. difficile* [2]. It remains an open question, to what extent, or if at all, asymptomatic carriers should be actively looked for in hospital settings, in long-term care facilities or before the initiation of antibiotic therapy. Endoscopic diagnosis of pseudomembranous colitis in the absence of other clinical diagnosis does not need laboratory confirmation of the presence of *C. difficile*.

The classical method to detect the free toxin directly from the stool sample by the cell culture cytotoxicity neutralization assay (CCCNA) has historically been the standard for the diagnosis of CDI. The other possibility is to isolate the *C. difficile* strain from the stool sample on selective media in an appropriate anaerobic environment and detect the toxin production of the isolated strain by any toxin detection method. This is referred to as toxigenic culture (TC) [2–5]. These reference methods are not considered practical; they need long turn-around time and special laboratory experience. CCCNA has for a long time been considered the "gold standard"; however compared to TC it has only a sensitivity of 75–85% mostly due to degradation of the toxin in the specimen during transport [3].

In most routine laboratories the *C. difficile*-specific antigens glutamate dehydrogenase (GDH) and toxins (TcdA and/or TcdB) are detected directly from feces, using enzyme immune assays (EIAs). EIAs show big variations in sensitivity and specificity compared to CCCNA or TC, ranging from 40% to 100% [3, 6]. EIAs



H. Sommermeyer, J. Piątek, *Clostridioides difficile*,
https://doi.org/10.1007/978-3-030-81100-6_6

have been demonstrated to have high negative predictive value, making them suitable for detecting negative samples [3].

Nucleic acid amplification tests (NAAT) to detect *C. difficile*-specific genes (16 rRNA, GDH) and the toxin genes (tcdA and tcdB) have been developed since the early 1990s. The detection of the binary toxin genes (cdtA and cdtB) and the 117 nucleotide deletion of the tcdC gene as a presumptive identification of ribotype 027 *C. difficile* is also possible by these methods [3]. NAATs specific for *C. difficile* toxins have a high negative predictive value; however, they detect the presence of the toxin genes and not the presence of the toxin proteins. This may give a positive result even in cases of asymptomatic carriers, leading to overdiagnosis and overtreatment of a patient. The use of this method as a stand-alone method is not recommended in diagnosing CDI [3–6].

The following recommendations have been proposed for an effective diagnosis of CDI [1]:

- Routine microbiology laboratories should carry out diagnostics for CDI from all unformed stool samples from patients >3 years old, without specific request from the physician.
- Formed stool samples should not be tested for CDI except in special situations (e.g., suspected ileus, when rectal swab is the recommended sample).
- Direct EIA for toxin detection or NAAT for toxin gene detection should not be used as stand-alone test.
- Either a two-step (Fig. 6.1) or a three-step algorithm (Fig. 6.2) for diagnosing CDI should be employed.
- Toxigenic culture and molecular typing should be carried out in an outbreak situation.
- Repeated testing after a positive result, as well as test of cure, is not recommended.

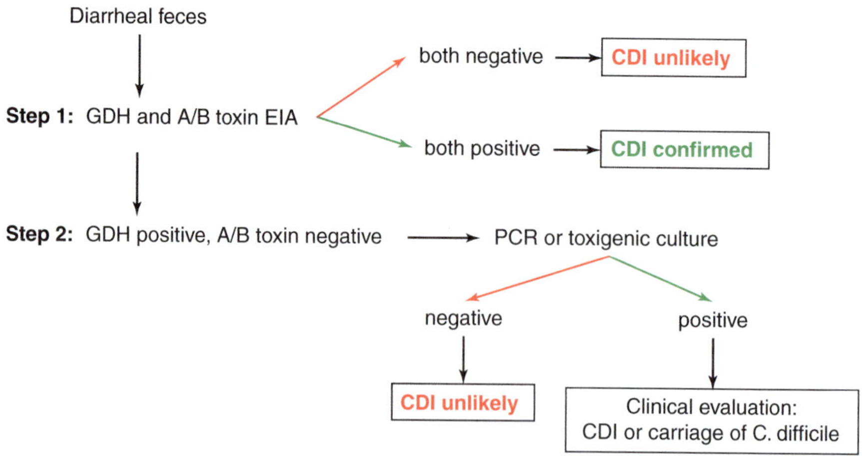


**Fig. 6.1**  Two-step algorithm for diagnosing CDI

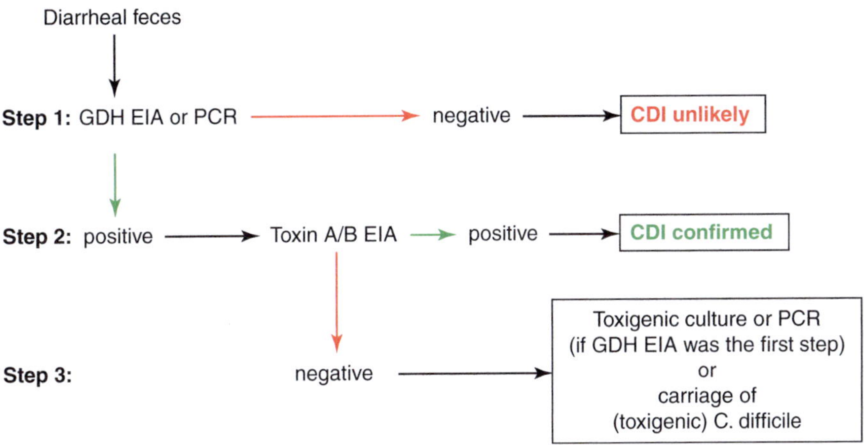

**Fig. 6.2** Three-step algorithm for diagnosing CDI

# References


1. Nagy E. What we know about the diagnosis, treatment and epidemiology of Clostridioides (Clostridium) difficile infection in Europe? J Infect Chemother. 2018;24:160–70. https://doi.org/10.1016/j.jiac.2017.12.003.
2. Bauer MP, Kuijper EJ, van Dissel JT. European society of clinical microbiology and infectious diseases (ESCMID): treatment guidance document for Clostridium difficile. Clin Microbiol Infect. 2009;15:1067e79.
3. Burnham CAD, Carroll KC. Diagnosis of Clostridium difficile infection: an ongoing conundrum for clinicians and for clinical laboratories. Clin Microbiol Rev. 2013;26:604e60.
4. Crobach MJ, Dekkers OM, Wilcox MH, Kuijper EJ. European society of clinical microbiology and infectious diseases (ESCMID): data review and recommendations for diagnosing Clostridium difficile-infection (CDI). Clin Microbiol Infect. 2009;15:1053e66.
5. Wilcox MH. Overcoming barriers to effective recognition and diagnosis of Clostridium difficile infection. Clin Microbiol Infect. 2012;18(6):13e20.
6. Crobach MJ, Planche T, Eckert C, Barbut F, Terveer EM, Dekkers OM, et al. European society of clinical microbiology and infectious diseases: update of the diagnostic guidance document for Clostridium difficile infection. Clin Microbiol Infect. 2016;22:S63e81.

## Contents



## 7.1    From Asymptomatic Carriers to Life-Threatening Presentations

The clinical picture of CDI is heterogeneous and ranges from the asymptomatic carrier state, mild or moderate diarrhea, to life-threatening pseudomembranous colitis (PMC). The incubation period is normally between 2 and 3 days; however in some cases it can also be longer than 3 days [1–4]. CDI can affect all parts of the colon, but the distal segment is most commonly affected. Most patients with CDI suffer from mild diarrhea and recover spontaneously after 5–10 days of antibiotic withdrawing. Diarrhea occurs in most cases during or directly after antimicrobial therapy, although CDI-caused diarrhea might also manifest a couple of weeks afterwards. The clinical features of CDI, in addition to watery diarrhea, include abdominal pain, fever, nausea and vomiting, weakness, and loss of appetite. Fecal occult blood test is often positive, although active bleeding is rarely present [1].

In the most severe clinical presentation of CDI, symptoms are life-threatening and include significant dehydration, abdominal distension, hypoalbuminemia with peripheral edema, and subsequent circulatory shock. Other severe complications of CDI include pseudomembranous colitis with toxic megacolon, colon perforation, intestinal paralysis, kidney failure, systemic inflammatory response syndrome, septicemia, and death [1].

Extracolonic manifestations of CDI are rare and most commonly involve small intestine infiltration, reactive arthritis, and bacteremia [5].

© The Author(s), under exclusive license to Springer Nature Switzerland AG 2021     65
H. Sommermeyer, J. Piątek, *Clostridioides difficile*,
https://doi.org/10.1007/978-3-030-81100-6_7

Mortality rate directly due to CDI is estimated at 5%, whereas mortality associated with CDI complications reaches 15–25%, and up to 34% in intensive care units (ICU). Mortality doubles in ICU patients with CDI, as compared with ICU patients without CDI [6–8]. Poor outcome is associated with older age, high leukocytosis, hypoalbuminemia, and high creatinine level [5, 9]. It has also been shown that the first-ever CDI episode increases the overall risk of death [6].

## 7.2 Pseudomembranous Colitis

Pseudomembranous colitis (PMC) is a manifestation of severe colonic disease. While CDI is the most common cause of PMC, it can be the result of a number of other underlying clinical conditions (Table 7.1).

**Table 7.1** Causes of pseudomembranous colitis

| Cause of pseudomembranous Colitis | Description | Comment |
| --- | --- | --- |
| Bacterial infection by *C. difficile* | Fulminant forms of infections | Most common cause of PMC, affecting 3–8% of CDI patients |
| Other bacterial infectious pathogens | *C. ramosum*, Enterohemorrhagic *Escherichia coli* O157:H7, *Klebsiella oxytoca*, *Plesiomonas shigelloides*, *Salmonella enterica*, *Shigella*, *Staphylococcus aureus*, *Yersinia enterocolitica* | PMC caused by these bacterial pathogens is often triggered by antibiotic therapy |
| Viral infectious pathogens | Cytomegalovirus | Mainly in AIDS patients |
| Parasitic infectious pathogens | *Entamoeba histolytica, Schistosoma mansoni, Strongyloides stercoralis* | |
| Colitis | Ischemic colitis | Insufficient blood flow in colonic vasculature resulting in ischemic damage |
| | Inflammatory bowel disease, Crohn's disease, ulcerative colitis | CDI in IBD patients is associated with an increased risk for PMC |
| | Microscopic colitis: Collagenous colitis, lymphocytic colitis | Grossly normal-appearing mucosa with abnormal biopsy findings of lymphocytic inflammatory infiltrate within the lamina propria, in 4–13% of patients with chronic diarrhea |
| | Behcet's disease | Systemic vasculitis that can involve small, medium, and large vessels |
| Medications | Alosetron (IBD treatment), cocaine, dextroamphetamine, gold, diclofenac, indomethacin, cisplatin, cyclosporine A, docetaxel, 5-fluorouracil | Most likely these agents cause PMC by localized ischemia and/or inflammation |
| Chemicals | Glutaraldehyde | Used for disinfecting and cleaning of endoscopes |

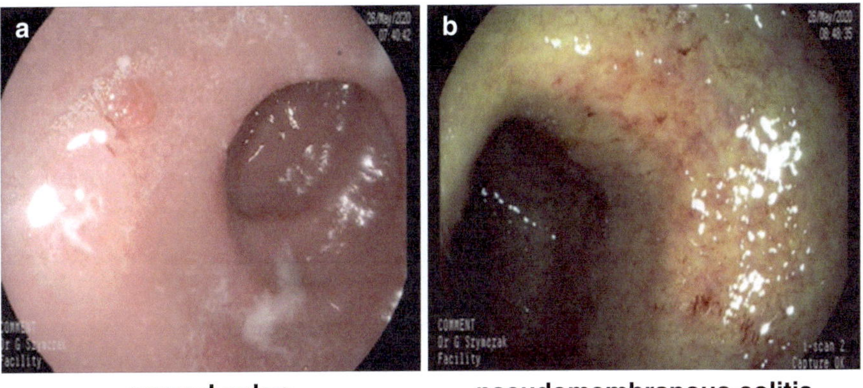


**Fig. 7.1** Endoscopic appearance of a normal (**a**) and PCM (**b**) colon

On endoscopic examination, PMC is characterized by elevated yellow-white nodules or plaques that form pseudomembranes on the mucosal surfaces of the colon (Fig. 7.1) [10, 11].

Endothelial damage from the initial event or disease process causes small areas of necrosis in the surface epithelium. The eruption of neutrophils, nuclear debris, and other inflammatory elements from the lamina propria onto the epithelium then leads to pseudomembrane formation [12, 13]. Pseudomembranes can be up to two centimeters in diameter, scattered among areas of normal or erythematous mucosa; however, confluent pseudomembranes that cover the entirety of the mucosa can be seen in severe disease [12, 14].

Classification of pseudomembranous lesions can be made based on the degree and depth of inflammatory changes, with grading of lesions from type 1 ("summit lesions," focal surface epithelial inflammation, or necrosis) to type 3 (complete mucosal necrosis and significant inflammatory debris) [14–16].

In fulminant CDI refractory to medical therapy or with complications (toxic megacolon, perforation with peritonitis, or septic shock), surgical intervention, including hemicolectomy or subtotal colectomy, may be necessary. Several retrospective cohort studies have found a survival benefit with early surgical intervention, particularly in patients undergoing total colectomy; nevertheless, overall morbidity and mortality in fulminant CDI is very high despite surgery, with mortality rates up to 80% reported [1, 17–20].

# References

1. McDonald LC, Gerding DN, Johnson S, Bakken JS, Carroll KC, Coffin SE, Dubberke ER, Garey KW, Gould CV, Kelly C, Loo V, Sammons JS, Sandora TJ, Wilcox MH. Clinical Practice Guidelines for Clostridium Difficile Infection in Adults and Children: 2017 Update by the Infectious Diseases Society of America (IDSA) and Society for Healthcare Epidemiology of America (SHEA). Clin Infect Dis. 2018;66(7):e1–e48. https://doi.org/10.1093/cid/cix1085.

2. McFarland LV, Mulligan ME, Kwok RY, Stamm WE. Nosocomial acquisition of Clostridium difficile infection. N Engl J Med. 1989;320:204–10. https://doi.org/10.1056/NEJM198901263200402.

3. Samore MH, DeGirolami PC, Tlucko A, Lichtenberg DA, Melvin ZA, Karchmer AW. Clostridium difficile colonization and diarrhea at a tertiary care hospital. Clin Infect Dis. 1994;18:181–7. https://doi.org/10.1093/clinids/18.2.181.

4. Toshniwal R, Silva J Jr, Fekety R, Kim KH. Studies on the epidemiology of colitis due to Clostridium difficile in hamsters. J Infect Dis. 1981;143:51–4. https://doi.org/10.1093/infdis/143.1.51.

5. Vaishnavi C. Clinical spectrum & pathogenesis of Clostridium difficile associated diseases. Indian J Med Res. 2010;131:487–99.

6. Czepiel J, Kędzierska J, Biesiada G, Perucki W, Nowak P, Garlicki A, Kedzierska J. Epidemiology of Clostridium difficile infection: results of a hospital-based study in Krakow, Poland. Epidemiol Infect. 2015;143:3235–43. https://doi.org/10.1017/S0950268815000552.

7. Sidler JA, Battegay M, Tschudin-Sutter S, Widmer AF, Weisser M. Enterococci, Clostridium difficile and ESBL producing bacteria: epidemiology, clinical impact and prevention in ICU patients. Swiss Med Wkly. 2014;144:14009. https://doi.org/10.4414/smw.2014.14009.

8. Vincent JL, Rello J, Marshall J, Silva E, Anzueto A, Martin CD, Moreno R, Lipman J, Gomersall C, Sakr Y, Reinhart K, EPIC II Group of Investigators. International study of the prevalence and outcomes of infection in intensive care units. J Am Med Assoc. 2009;302:2323–9. https://doi.org/10.1001/jama.2009.1754.

9. McFarland LV. Antibiotic-associated diarrhea: epidemiology, trends and treatment. Fut Microbiol. 2008;3:563–78. https://doi.org/10.2217/17460913.3.5.563.

10. Fekety R. Guidelines for the diagnosis and management of Clostridium difficile-associated diarrhea and colitis. Am J Gastroenterol. 1997;92:739–50.

11. Kawamoto S, Horton KM, Fishman EK. Pseudomembranous colitis: spectrum of imaging findings with clinical and pathologic correlation. Radiographics. 1999;19(4):887–97. https://doi.org/10.1148/radiographics.19.4.g99jl07887.

12. Carpenter HA, Talley NJ. The importance of clinicopathological correlation in the diagnosis of inflammatory conditions of the colon: histological patterns with clinical implications. Am J Gastroenterol. 2000;95(4):878–96. https://doi.org/10.1016/S0002-9270(00)00716-4.

13. Mylonakis E, Ryan ET, Calderwood SB. Clostridium difficile-associated diarrhea: a review. Arch Intern Med. 2001;161(4):525–33. https://doi.org/10.1001/archinte.161.4.525.

14. Price AB, Davies DR. Pseudomembranous colitis. J Clin Pathol. 1977;30(1):1–12. https://doi.org/10.1136/jcp.30.1.1.

15. Kelly CP, Pothoulakis C, LaMont JT. Clostridium difficile colitis. N Engl J Med. 1994;330:257. https://doi.org/10.1056/NEJM199401273300406.

16. Surawicz CM, McFarland LV. Pseudomembranous colitis: causes and cures. Digestion. 1999;60(2):91–100. https://doi.org/10.1159/000007633.

17. Koss K, Clark MA, Sanders DSA, Morton D, Keighley MRB, Goh J. The outcome of surgery in fulminant Clostridium difficile colitis. Color Dis. 2006;8(2):149–54. https://doi.org/10.1111/j.1463-1318.2005.00876.x.

18. Lamontagne F, Labbé AC, Haeck O, Lesur O, Lalancette M, Patino C, Leblanc M, Laverdière M, Pépin J. Impact of emergency colectomy on survival of patients with fulminant Clostridium difficile colitis during an epidemic caused by a hypervirulent strain. Ann Surg. 2007;245(2):267. https://doi.org/10.1097/01.sla.0000236628.79550.e5.

19. Longo WE, Mazuski JE, Virgo KS, Lee P, Bahadursingh AN, Johnson FE. Outcome after colectomy for Clostridium Difficile colitis. Dis Colon Rectum. 2004;47(10):1620–6. https://doi.org/10.1007/s10350-004-0672-2.

20. Surawicz CM, Brandt LJ, Binion DG, Ananthakrishnan AN, Curry SR, Gilligan PH, McFarland LV, Mellow M, Zuckerbraun BS. Guidelines for diagnosis, treatment, and prevention of Clostridium difficile infections. Am J Gastroenterol. 2013;108(4):478–98. https://doi.org/10.1038/ajg.2013.4.

## Contents



The European Society of Clinical Microbiology and Infectious Diseases (ESMID) has published two treatment guidelines in 2009 [1] and 2014 [2]. The guideline from the Infectious Diseases Society of America (IDSA) and the Society for Healthcare Epidemiology of America (SHEA) has been updated in 2018 [3]. These guidance documents were formulated by taking the results of published evidence-based studies about treatment options into account, including old (metronidazole, vancomycin) and new antibiotics (fidaxomicin, tigecycline), as well as alternative treatments (fecal transplantation, probiotics, or toxin-binding substances) [2].

## 8.1 Antibiotics-Based Therapy

### 8.1.1 Antibiotics Used for Treatment of CDI

Metronidazole (Fig. 8.1), introduced in 1960, is a nitroimidazole antibiotic used as antibacterial and antiprotozoal medication [4]. It is effective against infections caused by susceptible anaerobic organisms such as *Bacteroides*, *Fusobacterium*, *Peptostreptococcus*, *Prevotella*, and *Clostridium* [5]. Metronidazole inhibits the nucleic acid synthesis by disrupting the DNA of microbial cells.

© The Author(s), under exclusive license to Springer Nature Switzerland AG 2021
H. Sommermeyer, J. Piątek, *Clostridioides difficile*,
https://doi.org/10.1007/978-3-030-81100-6_8

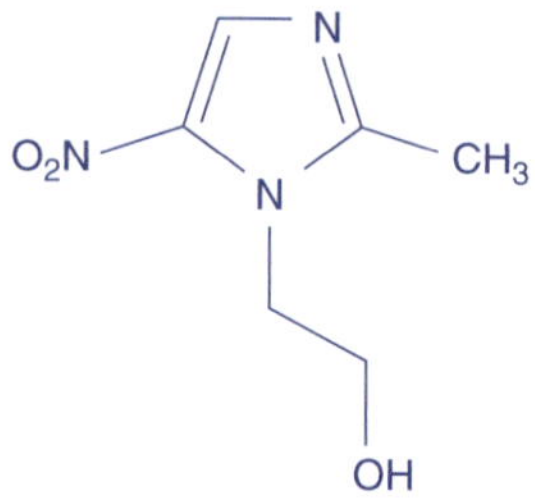

**Fig. 8.1** Chemical structure of metronidazole

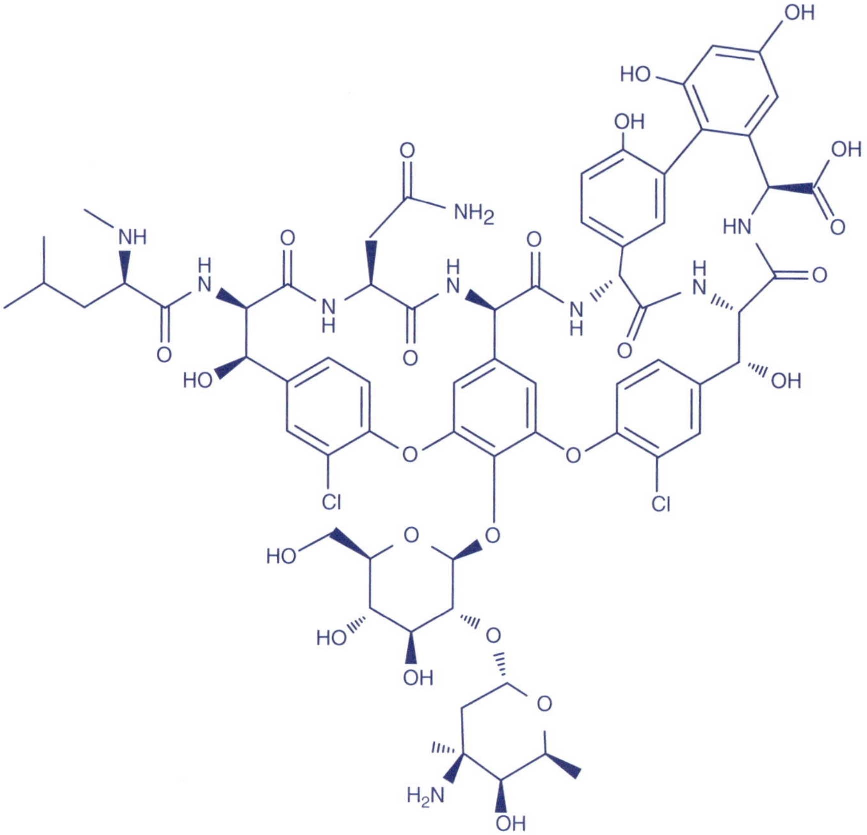

**Fig. 8.2** Chemical structure of vancomycin

Vancomycin (Fig. 8.2), introduced in 1954, is a glycopeptide antibiotic which is blocking the cell wall synthesis in Gram-positive bacteria [6]. It is given mostly intravenously, as it is not absorbed from the intestine. Vancomycin is used for the treatment of complicated skin infections, bloodstream infections, endocarditis, bone and joint infections, and meningitis caused by methicillin-resistant

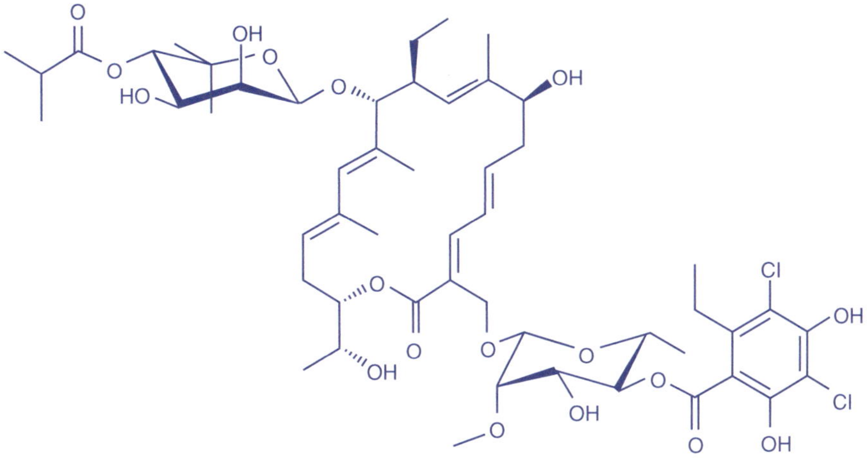

**Fig. 8.3**  Chemical structure of fidaxomicin

*Staphylococcus aureus*. Orally given as a treatment for severe *C. difficile* colitis, Vancomycin has the advantage of acting only in the intestine [5].

Fidaxomicin (Fig. 8.3), approved by the FDA in 2011, is a narrow-spectrum macrocyclic antibiotic acting against Gram-positive bacteria [7]. It is acting by the inhibition of the bacterial transcription (DNA → RNA). When given orally, fidaxomicin is hardly absorbed from the intestine. Fidaxomicin selectively eradicates pathogenic *C. difficile* with relatively little disruption to the multiple species of bacteria that make up the normal, healthy gut microbiota.

Clinical trials for approval showed that it is not inferior to oral vancomycin; clinical cure was achieved in 77.7% of patients (Vancomycin 67.1%), recurrence of CDI was lower (13.3%) than with oral vancomycin (24.0%) [8].

## 8.1.2  AB-Treatment of an Initial, Non-severe CDI Episode in Adults

The recommendation [3] for the treatment of an initial non-severe CDI episode comprises as a first step to discontinue therapy with the inciting antibiotic agent(s) as soon as possible, as this may influence the risk of CDI recurrence. The European Society of Clinical Microbiology and Infectious Diseases (ESCMID) recommends the administration of metronidazole (500 mg orally tid for 10 days) as a first-line drug, but indicates that this only applies to cases of mild infections [9]. According to the Infectious Diseases Society of America (IDSA) and the Society for Healthcare Epidemiology of America (SHEA) either vancomycin or fidaxomicin is recommended over metronidazole for an initial episode of CDI. The dosage is vancomycin 125 mg orally qid or fidaxomicin 200 mg bid for 10 days. As an alternative, non-antibiotic treatment regimen can also be considered (see below). In the USA, metronidazole 500 mg orally tid for 10 days is only suggested in settings where access to vancomycin or fidaxomicin is limited. Repeated or prolonged courses should be avoided due to the risk of cumulative and potentially irreversible neurotoxicity.

### 8.1.3  AB-Treatment of a Fulminant Episode of CDI in Adults

Fulminant CDI (previously referred to as severe, complicated CDI) is characterized by hypotension or shock, ileus, or megacolon. The treatment of choice [3] is administration of vancomycin. If ileus is present, vancomycin can be administered per rectum. The vancomycin dosage is 500 mg orally qid and 500 mg in approximately 100 ml normal saline per rectum every 6 h as a retention enema. Intravenously administered metronidazole should be administered together with oral or rectal vancomycin, particularly, if ileus is present. The metronidazole dosage is 500 mg intravenously every 8 h.

### 8.1.4  AB-Treatment of a First or Multiple Recurrence of CDI in Adults

Three regimes are recommended consideration for the treatment of a first recurrence of CDI [3], (i) oral vancomycin as tapered and pulsed regimen, (ii) 10-day course of fidaxomicin, or (iii) 10-day course of vancomycin in cases where metronidazole was used for the treatment of the primary episode.

Patients with >1 recurrence of CDI are recommended to be treated with (i) oral vancomycin using a tapered and pulsed regimen, (ii) a standard course of oral vancomycin followed by rifaximin, or (iii) fidaxomicin.

### 8.1.5  AB-Treatment of Non-severe Initial or First Recurrent CDI in Children

Either metronidazole or vancomycin is recommended for the treatment of children with an initial episode or first recurrence of non-severe CDI [3]. Dosage of metronidazole is orally 7.5 mg/kg/dose tid or qid for 10 days, with a maximum dose of 500 mg tid or qid. Vancomycin dosage is orally 10 mg/kg/dose qid for 10 days, with a maximum dose of 125 mg qid.

### 8.1.6  AB-Treatment of an Initial Episode of Severe CDI in Children

For children with an initial episode of severe CDI oral vancomycin is recommended over metronidazole [3]. Vancomycin (10 mg/kg/dose, qid) should be administered for 10 days (orally or per rectum) with or without metronidazole (7.5 mg/kg/dose tid or qid). Maximum doses for vancomycin and metronidazole are 125 mg qid or 500 mg tid or qid, respectively.

### 8.1.7   AB-Treatment of Second or Greater Episode of Recurrent CDI in Children

For children with a second or greater episode of recurrent CDI, oral vancomycin is recommended over metronidazole [3]. Administration of vancomycin should follow a tapered and pulsed regimen (10 mg/kg/dose, qid).

## 8.2   Fecal Microbiota Transplantation as Treatment for CDI

The fecal microbiota transplantation (FMT) [10] procedure is a non-antibiotic therapeutic approach to treat CDI which has received a lot of attention in the last decade. Interestingly enough, FMT has been known for over 1000 years, and was first described by a traditional Chinese medicine doctor, Ge Hong, who lived in the Dong Lin Dynasty period (284–364 BC). He applied human fecal suspension orally to patients with severe diarrhea or food poisoning [11]. In Europe the idea was first used in veterinary medicine by the Italian anatomist Fabricius Aquapendente in the seventeenth century [12]. Modern medicine first performed fecal transplantation in 1958 as a treatment for pseudomembranous enterocolitis (*C. difficile* had not been routinely identified at that time) [13]. The first report about fecal transplantation in a patient with confirmed *C. difficile* infection was published in 1983 [14].

A disturbed gut microbiota is considered a key factor in CDI development. Relative short antimicrobial therapy might dramatically reduce the amount of intestinal microbiota, but recovery may last several months. During that specific period patients lack their protective barrier (colonization resistance), and with exposure to spores an infection might develop quickly. Even though the gut microbiota is composed of thousands of species of microbes, it is thought that Bacteroides and Firmicutes play a predominant role in immunological responses against *C. difficile* [15].

Antibiotic withdrawal together with FMT has been demonstrated to have a high rate of prevention of recurrent CDI [16, 17]. After FMT patients showed increased fecal bacterial diversity, similar to that in healthy donors [17]. Transplantation with frozen fecal material has simplified the procedure [18]. Stool samples can be stored at −80 °C and used during the next 5–6 months. Fecal transplant can be administrated via oral capsules, lower gastrointestinal (GI) tract procedure (colonoscopy, retention enema), or upper GI tract procedure (nasojejunal/ nasoduodenal tube) [19]. Some potential complications of FMT are connected with delivery method (e.g., perforation with colonoscopy, aspiration pneumonia with upper GI administration). Safety concerns related to the transplantation of a more or less unknown highly complex mixture of microorganisms remain the main obstacle of FMT [20, 21]. Two key concerns associated with FMT are the risk of transferring infectious pathogens from the donor to the recipient and the development of autoimmunological disorders. The influence of gut microbiota on some immune-mediated diseases such as irritable bowel disease raises concerns about long-term side effects of fecal transplantation.

Nevertheless, FMT is recommended for patients with multiple recurrence of CDI who have failed appropriate antibiotic treatments [3].

## 8.3 Probiotics and Synbiotics as Treatment for CDI

A diverse and well-balanced gut microbiota is most likely the best protection against CDI. While FMT has been demonstrated to have beneficial effects at least in some patients, the safety concerns related to the administration of a highly complex and not fully characterized mixture of bacteria (and other components) is a clear obstacle of the FMT therapeutic approach. Alternatively, the administration of probiotics or synbiotics has been considered as they are supplementing the gut microbiota with well-characterized probiotic microorganisms. According to the 2001 definition by the World Health Organization (WHO), probiotics are "live microorganisms which, when administered in adequate amounts, confer a health benefit on the host." These probiotics must be alive when they are administered [22]. Probiotics can contain bacteria or yeast microorganisms. In the case of bacterial probiotics, mono-strain (containing only one strain of bacteria) or multi-strain (containing several bacterial strains) products have to be differentiated. Synbiotics contain in addition to the probiotic component(s) a prebiotic component which serves the probiotic organisms as a source of energy [23]. Commonly used prebiotics are fructooligosaccharides (FOS), galactooligosaccharides (GOS), or inulin. Mono-strain and multi-strain synbiotics have to be differentiated.

Probiotic bacteria express their beneficial effects at three levels. Firstly, they are limiting the growth of pathogenic microorganisms in the gut by a number of different mechanisms (Fig. 8.4) [24].

Secondly, they are involved in the regulation of the mechanic and biochemical barrier of the gut epithelium, by influencing mucus-producing Goblet cells, antibacterial peptide-secreting Paneth-cells and regulating tight junctions of the epithelium cells. At a third level, probiotics are involved in the maturing and regulation of the host immune system.

In a number of studies evidence was found that probiotics/synbiotics might play a role in the management CDI. In vitro-growth inhibition of *C. difficile* has been demonstrated for mono-strain [25, 26] and multi-strain probiotics or synbiotics [27]. Administration of *Lactobacillus acidophilus* and *B. bifidum* seemed to have a neutralizing effect on the toxins of *C. difficile*, as in a study it was found that only 46% of patients that received the probiotic were toxin-positive, compared to 78% of patients in the placebo group [28]. Colonization of *C. difficile* to epithelial cells could be prevented by administering a mixture of *Staphylococcus, Enterococcus, Lactobacillus, Anaerostipes, Bacteroidetes,* and *Enterorhabdus* [29]. The yeast probiotic *Saccharomyces boulardii* upregulated the expression of antitoxin A secretory

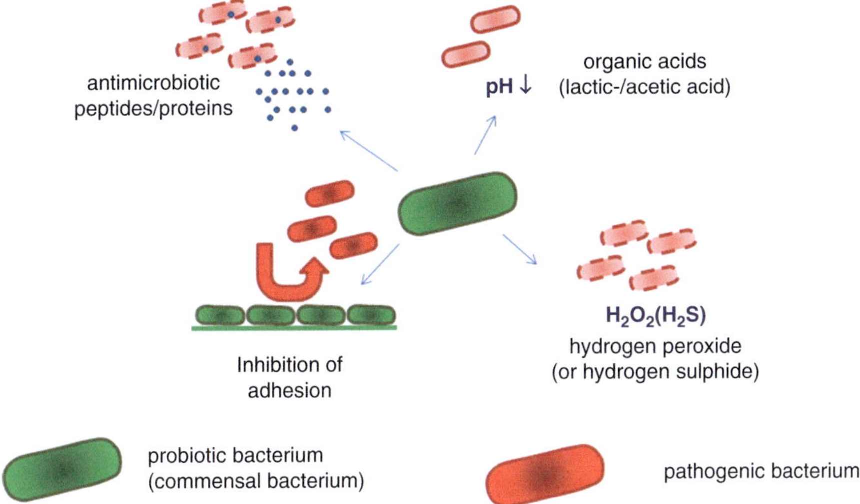


**Fig. 8.4** Probiotic mechanisms of action

immunoglobulin A expression in animal models of CDI and inhibited the binding of toxin A to epithelial cells [30, 31]. A mixed culture of nontoxigenic *C. difficile*, *Escherichia coli*, *B. bifidum*, and members of *Lachnospiraceae* was found to prevent the colonization of *C. difficile* in germ-free mice [32, 33].

In a meta-analysis study [34] it has been shown that probiotics are associated with a reduction in the incidence of CDI-associated diarrhea. This effect of probiotics was larger in children when compared with that in adults.

A review published in 2008 by the Cochrane Group [35] stated that there was not enough data to establish the role of probiotics for the treatment of CDI. In a more recent systematic review and meta-analysis by the Cochrane Collaboration, published in 2017, the authors came to the conclusion that probiotics have general positive effects in CDI patients [36]. Despite all the positive reports, the current clinical data are still not sufficient to result in a positive recommendation to use probiotics in the treatment of CDI. The main reason is that most of the clinical studies in which probiotics or synbiotics were investigated were not including a large enough number of patients.

## 8.4   Surgical Treatment of Complicated CDI

The term "complicated CDI" has been used to describe CDI cases with shock, megacolon, perforation, need for intensive care unit (ICU) admission, emergency colectomy, or death within 30 days of diagnosis [37]. Subtotal colectomy with end ileostomy is the procedure of choice for patients requiring surgery for severe colitis or toxic megacolon [38]. Subtotal colectomy allows for control of disease, stabilization of the patient and preserves reconstruction options for the future. Segmental colectomy has generally been associated with higher mortality and does not address the pan-colonic nature of the disease. Similarly, ileostomy alone, colostomy, and nontherapeutic laparotomy have generally shown increased mortality. An open approach is the standard of care in most severely ill patients. Careful mobilization of the abdominal colon is performed to prevent perforation and intraabdominal contamination. Vascular ligation near the colonic wall will minimize collateral injuries during a difficult dissection. In cases of severe colonic inflammation, ligating vessels first may minimize inflammatory response. Intraoperative colonoscopy is indicated if there is any question about the diagnosis.

The colon can be divided at the distal sigmoid or recto-sigmoid junction. Several options for the Hartmann pouch are possible. A longer Hartmann pouch can be brought up as a mucous fistula and matured at the bedside postoperatively or can be sewn to the inferior portion of the fascia. These methods for control of the stump help minimize complications resorting from a rectal stump blowout. Some surgeons advocate lavage of the rectal stump with vancomycin irrigation to help control inflammation in the Hartmann pouch. Data for this technique is limited. A Penrose drain positioned trans-anally may help to decompress the stump and prevent leak.

## References


1. Bauer MP, Kuijper EJ, van Dissel JT. European Society of Clinical Microbiology and Infectious Diseases (ESCMID): treatment guidance document for Clostridium difficile. Clin Microbiol Infect. 2009;15:1067e79.
2. Debast SB, Bauer MP, Kuijper EJ, on behalf of the Committee. European society of clinical microbiology and infectious diseases: update of the treatment guidance document for Clostridium difficile infection. Clin Microbiol Infect. 2014;20(Suppl 2):1e26. https://doi.org/10.1111/1469-0691.12418.
3. McDonald LC, Gerding DN, Johnson S, Bakken JS, Carroll KC, Coffin SE, Dubberke ER, Garey KW, Gould CV, Kelly C, Loo V, Sammons JS, Sandora TJ, Wilcox MH. Clinical Practice Guidelines for Clostridium Difficile Infection in Adults and Children: 2017 Update by the Infectious Diseases Society of America (IDSA) and Society for Healthcare Epidemiology of America (SHEA). Clin Infect Dis. 2018;66(7):e1–e48. https://doi.org/10.1093/cid/cix1085.
4. Metronidazole. https://www.drugs.com/metronidazole.html. Accessed 04/06/2021.
5. Zar FA, Bakkanagari SR, Moorthi KMLST, Davis MB. A comparison of vancomycin and metronidazole for the treatment of Clostridium Difficile-associated diarrhea, stratified by disease severity. Clin Infect Dis. 2007;45(3):302–7. https://doi.org/10.1086/519265.
6. Vancomycin capsules. https://www.drugs.com/cdi/vancomycin-capsules.html. Accessed 04/06/2021.
7. Fidaxomicin. https://www.drugs.com/dificid.html. Accessed 04/06/2021.
8. Louie TJ, Emery J, Krulicki W, Byrne B, Mah M. OPT-80 eliminates Clostridium difficile and is sparing of Bacteroides species during treatment of *C. difficile* infection. Antimicrob Agents Chemother. 2009;53(1):261–3. https://doi.org/10.1128/AAC.01443-07.
9. Coia JE, Kuijper EJ. The ESCMID study group for Clostridium difficile: history, role and perspectives. Adv Exp Med Biol. 2018;1050:245–54. https://doi.org/10.1007/978-3-319-72799-8_14.
10. Cammarota G, Ianiro G, Gasbarrini A. Fecal microbiota transplantation for the treatment of Clostridium difficile infection: a systematic review. J Clin Gastroenterol. 2014;48(8):693–702. https://doi.org/10.1097/MCG.0000000000000046.

11. Zhang F, Luo W, Fan Z, Ji G. Should we standardize the 1700 year old fecal microbiota transplantation? Am J Gastroenterol. 2012;107:755. https://doi.org/10.1038/ajg.2012.251.

12. Borody TJ, Warren EF, Leis S, Surace R, Ashman O, Siarakas S. Bacteriotherapy using fecal flora: toying with human motions. J Clin Gastroenterol. 2004;38:475–83. https://doi.org/10.1097/01.mcg.0000128988.13808.dc.

13. Eiseman B, SilenW BGS, Kauvar AJ. Fecal enema as an adjunct in the treatment of pseudomembranous enterocolitis. Surgery. 1958;44:854–9.

14. Schwan A, Sjolin S, Trottestam U, Aronsson B. Relapsing Clostridium difficile enterocolitis cured by rectal infusion of homologous faeces. Lancet. 1983;2:845. https://doi.org/10.1016/s0140-6736(83)90753-5.

15. Austin M, Mellow M, Tierney WM. Fecal microbiota transplantation in the treatment of Clostridium difficile infections. Am J Med. 2014;127:479–83.

16. Dutta SK, Girotra M, Garg S, Dutta A, von Rosenvinge EC, Maddox C, Song Y, Bartlett JG, Vinayek R, Fricke WF. Efficacy of combined jejunal and colonic fecal microbiota transplantation for recurrent Clostridium difficile infection. Clin Gastroenterol Hepatol. 2014;12:1572–6. https://doi.org/10.1016/j.cgh.2013.12.032.

17. van Nood E, Vrieze A, Nieuwdorp M, Fuentes S, Zoetendal EG, de Vos WM, Visser CE, Kuijper EJ, Bartelsman JFWM, Tijssen JGP, Speelman P, Dijkgraaf MGW, et al. Duodenal infusion of donor feces for recurrent Clostridium difficile. N Engl J Med. 2013;368:407–15. https://doi.org/10.1056/NEJMoa1205037.

18. Borody TJ, Khoruts A. Fecal microbiota transplantation and emerging applications. Nature Rev Gastroenterol Hepatol. 2012;9:88–96.

19. Bafeta A, Yavchitz A, Riveros C, Batista R, Ravaud P. Methods and reporting studies assessing fecal microbiota transplantation: a systematic review. Ann Intern Med. 2017;167:34–9. https://doi.org/10.7326/M16-2810.

20. Iqbal U, Anwar H, Karim MA. Safety and efficacy of encapsulated fecal microbiota transplantation for recurrent Clostridium Difficile infection: a systematic review. Eur J Gastroenterol Hepatol. 2018;30(7):730–4. https://doi.org/10.1097/MEG.0000000000001147.

21. Shogbesan O, Poudel DR, Victor S, Jehanger A, Fadahunsi O, Shogbesan G, Donato A. A systematic review of the efficacy and safety of fecal microbiota transplant for Clostridium difficile infection in immunocompromised patients. Can J Gastroenterol Hepatol. 2018;2018:1394379. https://doi.org/10.1155/2018/1394379.

22. Joint FAO/WHO Working Group. Report on Drafting Guidelines for the Evaluation of Probiotics in Food. London, Ontario, Canada, April 30 and May 1, 2002. https://www.who.int/foodsafety/fs_management/en/probiotic_guidelines.pdf. Accessed 04/06/2021.

23. Gibson GR, Hutkins R, Sanders ME, Prescott SL, Reimer RA, Salminen SJ, Scott K, Stanton C, Swanson KS, Cani PD, Verbeke K, Reid G. Expert consensus document: the international scientific association for probiotics and prebiotics (ISAPP) consensus statement on the definition and scope of prebiotics. Nat Rev Gastroenterol Hepatol. 2017;14(8):491–502. https://doi.org/10.1038/nrgastro.2017.75.

24. Bermudez-Brito M, Plaza-Díaz J, Muñoz-Quezada S, Gómez-Llorente C, Gil A. Probiotic mechanisms of action. Ann Nutr Metab. 2012;61(2):160–74. https://doi.org/10.1159/000342079.

25. Naaber P, Smidt I, Štšepetova J, Brilene T, Annuk H, Mikelsaar M. Inhibition of Clostridium difficile strains by intestinal Lactobacillus species. J Med Microbiol. 2004;53:551–3. https://doi.org/10.1099/jmm.0.45595-0.

26. Schoster A, Kokotovic B, Permin A, Pedersen PD, Dal Bello F, Guardabassi L. In vitro inhibition of Clostridium difficile and Clostridium perfringens by commercial probiotic strains. Anaerobe. 2013;20:36–41. https://doi.org/10.1016/j.anaerobe.2013.02.006.

27. Piatek J, Krauss H, Ciechelska-Rybarczyk A, Bernatek M, Wojtyla-Buciora P, Sommermeyer H. In-vitro growth inhibition of bacterial pathogens by probiotics and a synbiotic: product composition matters. Int J Environ Res Public Health. 2020;17(9):3332. https://doi.org/10.3390/ijerph17093332.

28. Plummer S, Weaver MA, Harris JC, Dee P, Hunter J. Clostridium difficile pilot study: effects of probiotic supplementation on the incidence of *C. difficile* diarrhea. Int Microbiol. 2004;7:59–62. https://doi.org/10.2436/im.v7i1.9445.

29. Lawley TD, Clare S, Walker AW, Stares MD, Connor TR, Raisen C, Goulding D, Rad R, Schreiober F, Brandt C, Deakin LJ, Pickard DJ, Duncan SH, Flint HJ, Clark TG, Parkhill J, Dougan G. Targeted restoration of the intestinal microbiota with a simple, defined bacteriotherapy resolves relapsing Clostridium difficile disease in mice. PLoS Pathog. 2012;8:e1002995. https://doi.org/10.1371/journal.ppat.1002995.

30. Buts JP, De KN, De RL. Saccharomyces boulardii enhances rat intestinal enzyme expression by endoluminal release of polyamines. Pediatr Res. 1994;36:522–7.

31. Qamar A, Aboudola S, Warny M, Michetti P, Pothoulakis C, LaMont JT, Kelly CP. Saccharomyces boulardii stimulates intestinal immunoglobulin A immune response to Clostridium difficile toxin A in mice. Infect Immun. 2001;69:2762–5. https://doi.org/10.1128/IAI.69.4.2762-2765.2001.

32. Corthier G, Dubos F, Raibaud P. Modulation of cytotoxin production by Clostridium difficile in the intestinal tracts of gnotobiotic mice inoculated with various human intestinal bacteria. Appl Environ Microbiol. 1985;49:250–2.

33. Reeves AE, Koenigsknecht MJ, Bergin IL, Young VB. Suppression of Clostridium difficile in the gastrointestinal tracts of germfree mice inoculated with a murine isolate from the family Lachnospiraceae. Infect Immun. 2012;80:3786–94. https://doi.org/10.1128/IAI.00647-12.

34. Lau CSM, Chamberlain RS. Probiotics are effective at preventing Clostridium difficile-associated diarrhea: a systematic review and meta-analysis. Int J Gen Med. 2016;9:27–37. https://doi.org/10.2147/IJGM.S98280.

35. Pillai A, Nelson R. Probiotics for treatment of Clostridium difficile-associated colitis in adults. Cochrane Database Syst Rev. 2008;1:CD004611. https://doi.org/10.1002/14651858.CD004611.pub2.

36. Goldenberg JZ, Yap C, Lytvyn L, Lo CKF, Beardsley J, Mertz D, Johnston BC. Probiotics for the prevention of Clostridium difficile-associated diarrhea in adults and children. Cochrane Database Syst Rev. 2017;(12):CD006095. https://doi.org/10.1002/14651858.CD006095.pub4.

37. Leclair MA, Allard C, Lesur O, Pépin J. Clostridium difficile infection in the intensive care unit. J Intensive Care Med. 2010;25(1):23–30. https://doi.org/10.1177/0885066609350871.

38. Seltman AK. Surgical management of Clostridium difficile colitis. Clin Colon Rectal Surg. 2012;25(4):204–9. https://doi.org/10.1055/s-0032-1329390.

## Contents



## 9.1   Probiotics and Synbiotics for Prophylaxis of CDI

As is the case with the role of probiotics and synbiotics in the treatment of CDI, there is also still a need for more data to conclude their potential role in prophylaxis of CDI [1]. Nevertheless, based on the currently available data and their excellent safety profile, pro- and synbiotics are frequently used by physicians for CDI prophylaxis. Most of the modern probiotics and synbiotics belong to the category of food supplements. Few of the currently marketed probiotic products are medical products and all of these medical products have been approved as such a number of decades ago. As technology has advanced, these old products, being trapped by their historical approval documentation, are no longer in-line with today's state-of-the-art requirements of probiotics and synbiotics.

Table 9.1 provides an overview about the product specification that should be fulfilled by a state-of-the-art synbiotic. While older products often fail to meet these requirements, more modern products are often designed in a way that most of the requirements are met. Probably the most important feature to differentiate among the myriad of products available is the protection of the probiotic bacteria against inactivation by the stomach acid. This is achieved for capsule products by special acid-resistance coatings of the capsule. Special coating procedures have been developed to protect powder products against the effects of low pH in the stomach. As the surface that has to be protected against an acid attack in powder products is very

**Table 9.1**  Product specification to be fulfilled by a modern synbiotic product

| Product feature | Required property | Reason(s) |
| --- | --- | --- |
| Number of different probiotics bacterial strains | 5–10 | A larger number of different probiotic strains can contribute a variety of mechanisms to contain pathogen growth in the gut, some of which might even act synergistically |
| Sufficient number of colony-forming units (CFU) | $\geq 1 \times 10^9$ | A sufficient high number of CFU is needed to result in colonization of the gut. However, this requirement is a function of other product features and probiotic strain characteristics |
| Protection against inactivation by stomach acid | Enteric coating | Enteric coating of capsules or powders will allow living bacteria to travel through low pH environment of the stomach |
| Synbiotic | Probiotic combined with prebiotic | The prebiotic component will provide a source of energy for the probiotic strains, thereby supporting the colonization of the gut by the probiotic bacteria |
| Allergens | Free of allergens | Product should be free of allergens, especially lactose-free and gluten-free |
| Capsule | Vegetarian capsule | Vegetarian capsules are gelatin-free and are acceptable for users who have to avoid products containing components sourced from pigs |
| Packaging | Blister or stick pack | Blister packaging or one-dose stick packs provide convenient administration |
| Dosing | Once daily | Once-daily dosing is supporting compliance of intake |
| Storage | At room temperature | Storage without refrigeration (needed for some older probiotics) simplifies logistics and storage by users |
| Acceptable price | Acceptable | As most pro- and synbiotics are not reimbursed by health insurance, price has to be low enough to allow out-of-pocket payment by patients |

large, special coating is needed, often using double coating layers, to achieve a sufficient protection of the probiotic bacteria in the powder products.

So far, the clinical data supporting a prophylactic usage of probiotics to prevent CDI is still not very strong. Nevertheless, patients with CDI risk factors or with a history of CDI are advised to take a probiotic or even better a synbiotic as a prophylactic measure to lower their risk of contracting CDI [2].

## 9.2   Bezlotoxumab for the Prevention of Recurrent CDI

Bezlotoxumab (Zinplava, Merck & Co.) is a human monoclonal antibody for the prevention (but not the treatment) of recurrent CDI in patients $\geq 18$ years of age receiving concomitant standard of care antibiotic therapy for CDI and who are at high risk for recurrence [3]. In two large randomized, double-blind, placebo controlled phase 3 trials, MODIFY I and II, patients who received bezlotoxumab in combination with standard of care therapy had a significantly lower rate of recurrent

CDI at 12-week follow-up compared with those who received standard of care therapy alone (MODIFY I: 17% vs. 28%; $P < 0.001$; MODIFY II: 15% vs. 26%; $P < 0.001$) [4]. Analysis of pooled trial data indicated that bezlotoxumab was particularly effective in patients with $\geq 1$ risk factor for recurrent CDI, with the greatest benefit in those with $\geq 3$ risk factors including age $\geq 65$ years, history of CDI in the previous 6 months, compromised immunity, severe CDI, and having a strain associated with poor outcomes of CDI [5]. A real-world multicenter study demonstrated prevention of recurrent CDI with bezlotoxumab comparable to clinical trial results regardless of type of standard of care treatment and timing of infusion. Multiple prior CDI recurrences were associated with a higher risk of subsequent recurrent CDI, supporting the use of bezlotoxumab earlier in the disease course [6]. As a monoclonal antibody-based therapy treatment, bezlotoxumab is expensive.

## References


1. Hickson M. Probiotics in the prevention of antibiotic associated diarrhoea and Clostridium difficile infection. Ther Adv Gastroenterol. 2011;4(3):185–97. https://doi.org/10.1177/1756283X11399115.
2. Issa I, Moucari R. Probiotics for antibiotic-associated diarrhea: do we have a verdict? World J Gastroenterol. 2014;20(47):17788–95. https://doi.org/10.3748/wjg.v20.i47.17788.
3. Bezlotoxumab (Zinplava). https://www.drugs.com/zinplava.html. Accessed 04/06/2021.
4. Wilcox MH, Gerding DN, Poxton IR, Kelly C, Nathan R, Birch T, Cornely OA, Rahav G, Bouza E, Lee C, Jenkin G, Jensen W, Kim YS, Yoshida J, Gabryelski L, Pedley A, Eves K, Tipping R, Guris D, Kartsonis N, Dorr MB, MODIFY I and MODIFY II Investigators. Bezlotoxumab for prevention of recurrent Clostridium difficile infection. N Engl J Med. 2017;376:305–17. https://doi.org/10.1056/NEJMoa1602615.
5. Gerding DN, Kelly CP, Rahav G, Lee C, Dubberke ER, Kumar PN, Yacyshyn B, Kao D, Eves K, Ellison MC, Hanson ME, Guris D, Dorr MB. Bezlotoxumab for prevention of recurrent Clostridium difficile infection in patients at increased risk for recurrence. Clin Infect Dis. 2018;67:649–56. https://doi.org/10.1093/cid/ciy171.
6. Hengel RL, Ritter TE, Nathan RV, Van Anglen LJ, Schroeder CP, Dillon RJ, Marcella SW, Garey KW. Real-world experience of bezlotoxumab for prevention of clostridioides difficile infection: a retrospective multicenter cohort study. Open Forum Infect Dis. 2020;7(4):ofaa097. https://doi.org/10.1093/ofid/ofaa097.

## Contents



Since its first isolation in 1935, *C. difficile* has developed into a major healthcare problem around the world. Over the last decades a tremendous amount of detailed knowledge about this microorganism and the pathogenic effects it can cause in humans has been accumulated.

## 10.1  Spores

As a spore-forming bacterium, *C. difficile* is difficult to eliminate even by aggressive hygiene measures. Therefore, it is of importance to be aware of potential sources of spores and engage in measures to limit, or better to contain the spread of spores. While it is obvious that a symptomatic CDI patient is a spore spreader, physicians and other healthcare providers have to keep in mind that spores can also be spread by asymptomatic *C. difficile* carriers. The percentage of adults being asymptomatic *C. difficile* carriers varies significantly, but is highest in those with a history of previous healthcare service contacts (e.g., hospital stays). As a high percentage of children below 2 years of age are frequently asymptomatic carriers of *C. difficile*,



H. Sommermeyer, J. Piątek, *Clostridioides difficile*,
https://doi.org/10.1007/978-3-030-81100-6_10

they have also to be considered as a potential source of *C. difficile* spores. Being aware of potential sources of *C. difficile* spores is what allows implementation of the appropriate measures (i) to reduce or eliminate spores by hygiene measures and (ii) to limit or prevent the risk of transmission from *C. difficile* infected patients to others.

## 10.2   Pathophysiology

The pathophysiology of *C. difficile* has been extensively characterized in recent years and an in-depth understanding of the central role of the *C. difficile* toxins has been established. While research is still needed to resolve some aspects, a very detailed insight into the molecular processes by which *C. difficile* is executing its sometimes devastating pathogenic effects is available. While scientifically fascinating, these insights should always remind us that *C. difficile* is a potentially deadly bacterial pathogen.

## 10.3   Epidemiology

Epidemiological studies have revealed that *C. difficile* and its different strains (e.g., ribotype 027) have evolved into a deadly threat by human activities. While the invention of antibiotics can be considered as one of the most important medical inventions of humankind, the broad usage of antibiotic therapy has also been a key driver for making *C. difficile* a growing major problem for healthcare. All oral antibiotic therapy has also a major influence on the gut microbiota. Especially broad-spectrum antibiotics can eliminate the "colonization resistance" provided by a diverse and well-balanced gut microbiota. Disturbing the gut microbiota by antibiotics always carries the risk of uncontrolled growth of pathogens (e.g., *C. difficile*) present in the gut. The broad usage of antibiotics has not only made the *C. difficile* problem bigger but has also created a more severe quality of the problem. After the introduction of new 8-methoxyquinolone antibiotics (gatifloxacin and moxifloxacin) a new hypervirulent *C. difficile* strain (ribotype 027) has emerged. CDI with this particular strain of *C. difficile* is associated with high mortality rates due to the production of large amounts of toxins.

## 10.4   Risk Factors

The key risk factors of initial CDI have been identified and well characterized in recent years. Top risk factors are (i) exposure to healthcare, (ii) antibiotic therapy, (iii) proton pump inhibitor therapy, (iv) advanced age, and (v) cancer therapy. Knowing the risk factors is of importance for two reasons. Firstly, knowledge of the risk factors can guide some general medical behaviors (e.g., careful usage of antibiotics and proton pump inhibitors). Secondly, it allows physicians to tailor patient management according to the respective risk profile of this particular patient. Overutilization of certain medications has been identified in a lot of cases indicating that a rational usage of antibiotics and proton pump inhibitors should be high on the agenda of every physician.

The primary risk factor of recurrent CDI is a previous infection with *C. difficile*. While this indicates the importance of avoiding an initial CDI, this risk factor is of importance for the future management of CDI patients.

## 10.5 Diagnosis

Diagnostic procedures for detecting *C. difficile* are well established. It remains an open question to what extent, or if at all, asymptomatic carriers should be actively looked for in hospital settings, in long-term care facilities or before the initiation of antibiotic therapy.

## 10.6 Clinical Picture

CDI can present itself in a broad range of clinical pictures ranging from mild diarrhea to life-threatening pseudomembranous colitis. Most cases of pseudomembranous colitis are caused by CDI; however, physicians should be aware of the fact that it could also result from other causes.

## 10.7 Treatment

At first glance it might look paradoxical that infections caused by the bacterial pathogen *C. difficile* are mainly caused by antibiotic therapy. With today's knowledge however it is clear that a first step in the management of CDI is the termination of the inflicting antibiotic therapy. Current antibiotic therapy of CDI is focused on oral therapy with either vancomycin or fidaxomicin. However, these two antibiotics have also been described to cause fulminant colitis. As antibiotic therapy is one of the dominant risk factors for CDI, interest in alternative therapeutic approaches has emerged. Fecal microbiota transplantation has been demonstrated to be effective in preventing recurrent CDI. However, administration of a not fully characterized mixture of microorganisms will always remain a major concern related to this therapeutic approach. Probiotics or synbiotics provide a more defined microbiological treatment approach, but to finally judge their potential therapeutic effectiveness more clinical data are still needed.

## 10.8   Prophylaxis

Probiotics and synbiotics have been demonstrated to have beneficial effects in patients suffering from CDI-related diarrhea. Based on their general good safety profile and low price, probiotics and synbiotics can be considered for prophylactic treatments, at least in patients with certain CDI risk profiles. While the use of probiotics or synbiotics has still not made it into the official guidelines for prophylaxis of CDI, data indicate that they are good candidates with an excellent safety profile.

The monoclonal antibody bezlotoxumab has been approved for prevention of recurrent CDI in certain patients; however, high treatment costs most likely will limit broader usage.

## 10.9   General Outlook

The attention for CDI by healthcare and the broader public is fluctuating over time. Every deadly outbreak of *C. difficile* can push this topic to the front-page of the news. However, in our modern times, interest might be high for a moment but will also vanish fast. For physicians and other healthcare professionals it is of utmost importance to be aware that the problem of managing CDI is always present and will not go away. It is in their hands to control the problem to the best possible extent. It has to be kept in mind: ***"Good knowledge of a problem can strongly contribute to a successful management of a problem."***